An Observer's Guide to the

Glaciers of Prince William Sound, Alaska

By Nancy R. Lethcoe
Preface by William O. Field
Illustrations by R. James Lethcoe

Prince William Sound Books: Valdez, Alaska.

The following publishers have generously given permission to use illustrations from copyrighted works: From *EARTH* by Frank Press and Raymond Siever. Copyright © 1974, 1978, 1982, 1986 W.H. Freeman and Company. Reprinted with the permission of W.H. Freeman and Company. From *Glacial and Fluvioglacial Landforms* by R.J. Price. Copyright © 1972 Longman House. Reprinted with the permission of Longman House. From "Holocene Vegetation History of the Prince William Sound Region, South-Central Alaska" by Calvin J. Heusser in *Quaternary Research* 19: 352. Copyright © The University of Washington. Reprinted with the permission of the Unviersity of Washington. From *A.G.S. Glacier Studies* maps by William O. Field "Glacier Termini of Icy Bay" (No. 64-4-G5), "Survey and Photographic Stations in Harriman Fiord and Barry Arm, Prince William Sound, 1964" (No. 64-3-G9), "Survey and Photographic Stations in Upper College Fiord, Prince William Sound Alaska 1964" (No. 64-3-G10). Reprinted with the permission of William O. Field. Nancy Simmerman has graciously allowed me to use her copyrighted photographs.

© 1987 by Nancy R. Lethcoe.
All rights reserved. No part of this book may be reproduced in any form or by any electronic or mechanical means, including information storage and retrieval systems, without permission in writing from the publisher.

Cover photo: Columbia Glacier. 1976. Austin Post.

ISBN 0-9613146-6-4

Acknowledgements

I gratefully acknowledge the assistance of Austin Post, William O. Field and Kristine Crossen in providing information and previously unpublished material and Nancy Simmerman in providing photographs and professional services. Special thanks is due to Calvin Hazelet for permission to quote from George C. Hazelet's diary. I hope the diary, itself, will soon be published. I wish also to express my appreciation to William O. Field, Edward R. LaChapelle, Bruce Molnia, and Austin Post for their careful reading of all or parts of the manuscript. All errors are, of course, the author's. In the preparation of the manuscript, I thank Barbara and Peter Sokolov, Lane Mayer, Lisa Cotter, and my husband. For the many other ways in which I received special assistance, I would like to acknowledge the help of John and Lisa Cotter, John and Dolores Crowley, Nancy Mueller, Stan Stephens, Tim Jones, Jerry Massman, Lawrence Nielsen, Glenn Juday, Joe Kurtak, John Weiland, and Marjorie Slatterly. Finally, I would like to thank my students at Prince William Sound Community College and our charter guests for the many excellent questions they asked — questions that stimulated me to seek ever further for the answers.

It would not have been possible to write this book in a small town like Valdez were it not for the considerable financial support the Valdez City Council has given to the Valdez Consortium Library and the Valdez Heritage Center/Museum. Thank you. It is also a pleasure to thank the librarians and the Heritage Center staff for their encouragement and assistance.

Nancy R. Lethcoe

Table of Contents

Acknowledgements	iii
Preface	v
1. Introduction	1
2. Glacial Formation and Movement	8
3. Types of Glaciers	17
4. Glacial Features	22
5. Glacial Landscapes	33
6. Glacial Drift	39
7. Preliminary History of Prince William Sound's Glaciation	43
8. Exporation and Scientific Investigation	50
9. Retreat: Coastal Glaciers of Passage Canal and the Kenai Mountains	56
10. Of Gentlemen and Glaciers. Port Wells, Barry Arm, and Harriman Fiord	80
11. Of Advance and Retreat. College Fiord	98
12. A Tale of Two Glaciers. Meares and Columbia	111
13. On the Trail of the Prospectors. Port Valdez area	124
14. Drama at the Bridge. Cordova Area	138
Bibliography	146
Index	149

Preface

The fortunate people who have visited Prince William Sound and those who will be privileged to do so in the future will welcome this latest book on an unusually beautiful coastal area of South Central Alaska. Glaciers amidst snow-capped mountains reaching to over 13,000 feet, the rocky crags merging down-slope to grassy uplands, and on to dense forests which line the shores of sparkling inlets create a scene of unsurpassed beauty. The glaciers become much more intriguing when one is aware of their history as reported by early explorers and scientists.

This is a special book by an author familiar with both the pre-history and recorded information of the past ninety odd years.

The glaciers of Prince William Sound have a special charm and interest. Almost all types of glaciers are to be found among the dozen or so fiords that lead from the Sound into the heart of the Chugach and Kenai Mountains. Each glacier has a personality, often very different from that of its neighbors. To one who, after several visits becomes familiar with the area, individual glaciers become old friends, whose behavior is a matter of concern. Are they shrinking, growing, or more or less holding their own?

The author of this book has travelled the inlets of Prince William Sound and seen and photographed its many glaciers under a variety of conditions. Equally important, she is familiar with the early literature and much of the recent research, so that her book is an up-to-date discussion of observations in such fields as geology, geo-botany, and glaciology.

Few observations of glaciers were made prior to the visit of the Harriman Alaska Expedition in June, 1899. It brought together an unusual team of eminent scientists representing many disciplines, provided ideal conditions for field observations, and then produced a series of publications of the highest quality. The classic report by Grove Karl Gilbert of the U.S. Geological Survey on the glaciers and his concept of glaciation forms Volume Three of the Harriman Series. For the northern inlets of Prince William Sound, it was the beginning of careful, systematic observations and detailed mapping. Many of the picture viewpoints and survey stations established by him and geographer and mapmaker, Henry Gannett, are still in use.

A few years into this century, 1905-1909, the Geological Survey's team of U.S. Grant and D.F. Higgins completed the examination, photography, and mapping of all the remaining inlets of the Sound. Their report was first published in the *Bulletin of the American Geographical Society* in 1910-1911, and then again as *USGS Bulletin 526* in 1913. At the same time, the glaciers were also being studied by Professors Ralph S. Tarr of Cornell University and Lawrence Martin of the University of Wisconsin, whose studies from 1909 to 1911 were supported by The National Geographic Society and resulted in the monograph *Alaskan Glacier Studies*, published in 1914 by that Society.

Thus all the principal glaciers have been known and named for over 75 years, and the changes they have undergone have been documented and recorded in surveys, texts, and photographs.

With the introduction of aerial photography in the 1930s, an entirely new dimension was introduced. Major changes could easily be detected and the upper reaches of each of the glaciers, previously seen only near their termini, were revealed. This did not take the place of observations from ground level, but provided a new perspective and more extensive information. The latest tool for the glaciologist is satellite imagery, developed in the 1970s and still undergoing improvement and refinement. Despite these new techniques and the invaluable work of early researchers, there is a continuing need to record the changes in the glaciers whenever an opportunity presents itself. Nancy Lethcoe's book provides the necessary information of sites of the most advantageous photo stations or viewpoints from which pictures can be taken for comparison with existing records. Such pictures reveal a great deal more than just changes in the position of glacier termini. They also can show changes in the general appearance and character of the glaciers, in the vegetation on nearby ice-free areas, and in whatever evidence there may be of shrinking or growth in the upper parts of the glaciers.

The valuable role that persons who are not scientists can play was realized by Harry Fielding Reid of John Hopkins University who wrote to the Editor of *Science* in a letter of 23 May 1896 and published in the issue of 23 June: "Many of your readers will doubtless visit American glaciers this summer, either on the Pacific Coast, or Canada, or in Alaska; and I hope they will take sufficient interest in the subject to make observations which will be of value. The information most desired regarding any glacier is whether it is advancing or retreating." A few years later G.K. Gilbert wrote: "For the study of changes in the size of glaciers, photographic views are of peculiar value. A view showing a glacier in relation to details of adjacent land constitutes a record which can at any time be compared either with the objects themselves or with another photograph made in another year or month." He added a note emphasizing the importance of recording the year, month, and day a photo was taken, and indicating where prints and negatives are filed.

It should be pointed out that while the big glaciers tend to be the most spectacular, the small glaciers are equally interesting from a scientific standpoint. This is because while all glaciers are sensitive to a change in local meteorological factors, the small glaciers respond much more quickly than the large ones, and provide more immediate clues of change in the local climate. While a large glacier responds to these same influences, this may not be evident for decades.

Unlike the glaciers of most other areas in Alaska and other parts of the world, most of the glaciers of Prince William Sound have not yet receded very far from their maximum positions of recent centuries. Here one can still see active glacial ice edging into trees or other forms of vegetation several hundred years old, indicating that they have not been farther forward in that period of time. This is a sight which is exceptional in the present world. Other glaciers are receding slowly from moraines formed as late as the last quarter of the 19th century. There are reasons for these anomalies. But what are they? One hopes that scientists will discover the answers by the next century, but in the meantime, every bit of evidence may be useful in working out a satisfactory solution to the riddle.

Much of our information about the behavior of the glaciers in Prince William Sound has come from photographs taken by both scientists and others who have visited the area over the past century. Nancy Lethcoe's book provides the background information on the history of change at each of the principal glaciers. A visitor who wishes to contribute to the record could do so by taking pictures from the established sites and making their photos available to the author for safe keeping and eventual transmission to appropriate archives. Such a contribution would be a great value for further analysis of the changes which are taking place in the Prince William Sound area.

This book provides information on where to go in Prince William Sound to see the glaciers and to make scientifically useful observations, which, perhaps unsuspected at the time, may provide an important clue in recording and analyzing glacier behavior. These data may in turn provide evidence which can be interpreted in determining information leading to an explanation of some of the complex changes which may ultimately affect the survival of life as we know it on Planet Earth.

William O. Field
September 22, 1987

Chapter 1. Introduction:

LOCATION: Prince William Sound straddles the 60th parallel and is farther north than Cape Horn is south. Within its 10,000 square miles are 3,000 miles of heavily convoluted coastline. To the south, Montague Island serves as a fifty mile long breakwater protecting the sound from the often tempestuous Gulf of Alaska. To the west, north, and east, the high peaks and vast icefields of the Kenai and Chugach mountains create an effective barrier geographically isolating the sound from adjacent areas.

Fig-1.

ACCESS: Three small towns lying along the sound's perimeter provide access to the region. Glaciers on the western side of the sound are best approached from the port of Whittier, sixty railroad miles southeast of Anchorage. From this isolated community, charter boat operators provide access to the glaciers of Port Nellie Juan, Icy Bay, Barry Arm, Harriman Fiord, College Fiord, and Unakwik Inlet. A tourboat provides excursions to College and Harriman Fiords. These are all within the proposed Nellie Juan and College Fiord/Columbia Glacier Wilderness Areas. Several glaciers and many glacial features can be seen along the ferry and tour boat route between Whittier and Valdez.

Meares, Columbia, Shoup, and Valdez Glaciers are reached from the port of Valdez, one of the most beautiful towns in Alaska, which lies in the northeastern corner of the sound. Valdez is connected to the rest of the state by the Richardson Highway. From here tourboats make calls at Columbia or Shoup Glaciers. Outfitters in Valdez offer guided mountaineering trips onto the glaciers and plane or helicopter flightseeing excursions.

From the picturesque, but isolated, fishing community of Cordova on the southeastern corner of the sound, one can visit glaciers along the Copper River Highway. Charter boats from Cordova also regularly visit the glaciers of the northern sound. Rental cars are available for those wishing to drive out the Copper River Highway to see Sheridan (Mile 12), Childs (Mile 48) and Miles (Mile 48) Glaciers. Plane and helicopter flightseeing is also available. The most exciting and interesting way to view Miles and Childs Glaciers plus several others is to take a raft trip down the Copper River from Chitina. Several companies offer this service. Other glaciers along the Copper River are not covered in this book because they are outside of the Prince William Sound area.

In addition to the state ferry, cruise ships, and large tourboats, numerous small charter companies using powerboats, rafts, sailboats and kayaks all offer trips. Some companies feature special glacier and naturalist-guided trips. For additional information on businesses providing recreational opportunities in the Prince William Sound region write the Alaska State Division of Tourism, P.O. Box E-001, Juneau, AK 99811.

GEOLOGY: The geology of the Prince William Sound region and Alaska is only recently being unraveled, and many perplexing problems concerning the origins of the area remain. Alaska is composed of many terranes, fragments of former continents and oceanic plates, that moved northward at various times and eventually joined together. The Chugach/PrinceWilliam Terrane is the youngest terrane in Alaska. The presence of pillow basalts, which are formed when magma cools under water; the paleomagnetic dating of the rocks; and massive bands of graywacke, which are rocks formed from sediments deposited along a submarine canyon wall lead some geologists to suspect that the Prince William Terrane originated along the edges of an ancient mid-oceanic ridge-rift system (hence the pillow basalts) located at about 25° North suggested by paleomagnetic dating) that was being subducted under the continental plate (hence the graywackes). The Chugach/PrinceWilliam Terrane collided with coastal Alaska about 40 mya (million years ago). Between 60 and 40 mya, granitic batholiths bulged up from the earth's interior uplifting the overlying sedimentary rocks without breaking the surface. These granitic batholiths, occurring throughout the sound, may have been associated with forces released during the accretionary process. Since then, the landscape has been shaped by earthquakes and extensive glaciation, which probably removed thousands of feet of sedimentary rock. It was the persistent erosive power of these glaciers that has in places laid bare the underlying granites and carved out the sound's complex system of fiords. Today, the twin forces of earthquakes and glacial erosion continue to act out their eternal dance of creation and destruction.

CLIMATE: The sound has a maritime climate with moderate temperatures and copious precipitation. To the south, Kodiak Island experiences a warmer oceanic climate while to the north interior Alaska has a continental type climate typified by drier weather and

Fig-2. The Alaska State ferry, *E.L. Bartlett*, provides car and foot passenger service between Whittier, Valdez, and Cordova. Seth Glacier (Passage Canal) is in the background. Note the light colored rock on the ridge to the right where a granitic batholith is emerging from beneath the retreating glacier. Photo by author, 1982.

extreme temperature ranges. Although precipitation may fall as rain most of the year at sea level, the moist warm air soon cools as it rises over the steep mountains where it drops prodigious amounts of snow — 400 to 800 inches annually.

Because of its varied terrain, the sound has many microclimates — regions where the climate differs significantly from adjacent areas. For example, often on a clearing day following a period of inclement weather, rain continues to fall over forested hillsides while the area around a glacier is light and sunny. Rainfall persists over the forested areas because moisture evaporating from warm, vegetated land rises, cools and releases precipitation again. Meanwhile, the glaciers lose much less water to evaporation and

Fig-3. Glaciers have their own locally generated microclimates. Note the rising evaporation clouds and rain falling over the forested hillsides while the area above Harvard Glacier (College Fiord) is clear. Photo by author.

reflect light (albedo effect). Consequently, the clouds are fewer and appear lighter near a glacier. Sailors are familiar with another microclimatic condition associated with the glaciers. Because of their colder temperatures, local high pressure systems form over the glaciers while the adjacent sound, with its warmer water, tends to have a lower pressure over it. The difference in pressure causes air to move from the area of higher pressure to that of lower pressure creating local glacier breezes.

Early explorers, whose brief visits coincided with gales, described the sound as wet, cold, foggy and generally inhospitable. Those who saw the sound in the sunshine found it an unforgettable land whose chief resource was its unparalleled scenic quality. John Muir was one of the latter:

> But just as we entered the famous Prince William Sound, that I had so long hoped to see, the sky cleared, disclosing to the westward one of the richest, most glorious mountain landscapes I ever beheld — peak over peak dipping deep in the sky, a thousand of them, icy and shining, rising higher, higher, beyond and yet beyond one another, burning bright in the afternoon light, purple cloud-bars above them, purple shadows in the hollows, and great breadths of sun-spangled, ice-dotted waters in front. The nightless days circled away while we gazed and studied, sailing among the islands, exploring the long fiords, climbing moraines and glaciers and hills clad in blooming heather — grandeur and beauty in a thousand forms awaiting us at every turn in this bright and spacious wonderland. But that first broad, far-reaching view in celestial light was the best of all. (Muir 1901, I:132).

Today, what astounds many people most about the sound is the presence of so many glaciers at sea level where the temperatures are obviously above freezing during the summer and for much of the winter as well. Their first thought is often — why don't the glaciers just melt? In general, the answer, of course, is that the glacial ice is melting, but as fast as it melts, new ice flows down the mountainside to replace it.

PRINCE WILLIAM SOUND'S GLACIERS: Prince William Sound is noted for its surrounding icefields from which flow numerous glaciers that often reach sea level. Contrary to popular misconceptions, the sound does not contain its innumerable glaciers just because of cold temperatures. Winter temperatures in interior Alaska, where there are fewer glaciers, are much colder. Instead, so much snow falls during the winter months on the Chugach and Kenai mountains that it cannot all melt during the summer. Gradually, glaciers form and begin to move under the force of gravity toward the sea.

Prince William Sound's glaciers are classic examples of temperate glaciers. Unlike polar glaciers, such as those found in Greenland and Antarctica, which are below freezing and static, temperate glaciers are at or near the freezing point during most of the year and slide over their bedrock. Because snow recrystallizes into ice much faster when it thaws and refreezes repeatedly, the time it takes snow to become glacial ice is much shorter on a temperate glacier than on a polar one.

The Prince William Sound region, with seventeen tidewater (calving) glaciers, is unique in having so many glaciers terminating at sea level. Topography plays an important role in the large number of tidewater glaciers. For many of these, the descent from their accumulation areas to tideline is extremely abrupt.

Until recently, all but one of the tidewater glaciers has undergone a major retreat within the last few centuries. Some, like Columbia, Nellie Juan, Yale and Shoup, are in the drastic retreat phase of their cycle. Other formerly tidewater glaciers now terminate on land, such as Falling and Taylor glaciers. Nine tidewater glaciers are now slowly advancing: Blackstone, Harriman, Surprise, Cascade, Coxe, Bryn Mawr, Wellseley, Harvard and Meares.

HOW TO USE THIS BOOK: This book is designed both for those who view the glaciers from a distance and for those who have the opportunity to go ashore for more detailed exploration. Chapters 1 to 6 explain observable glacial phenomena and give locations where they may be readily seen. Almost every feature is illustrated by a photograph or drawing. Chapters 7 to 12 discuss the history of Prince William Sound's glaciation and provide detailed information on the recent history, exploration, and unique features of forty-five major glaciers. Both historical and recent photographs illustrate the text.

GLACIER OBSERVATION: Every year, tens of thousands of visitors come to Prince William Sound to watch the glaciers. For some, it is a once in a lifetime experience. For others, it is a yearly ritual, an exciting time to see what the glaciers are doing this year.

Why do these seemingly immobile masses of ice yield such a mysterious, yet pervasive appeal? Perhaps it is our awareness of the extreme mutability underlying a beguiling appearance of changelessness that explains our fascination for glaciers. Or maybe it is the primeval attraction to power that draws people. For here is a natural force that man does not control. However much we may wish to speed up a glacier's action of grinding down bedrock into the mineral component of fertile soil, we are helpless to do so. However much we may wish to stop a glacier's inexorable advance across the mouth of a mine or down a fiord, we cannot do so. Certainly, for many people glaciers, like few other geological phenomena, stimulate strong feelings of empathy. They feel joy when a glacier, such as Harriman, repairs the embayment that threatened the safety of its terminus' position and a sense of loss, even grief, when Columbia fails to do so and starts to retreat, giving up the ground it had taken perhaps 3,000 years to gain, inch by incredible inch. For others, the mysticism of history attracts them to the glaciers. These remnants of ice still inhabit the same valleys from which their ancestors, the great ice age glaciers, poured forth to cover the land and carve out Prince William Sound. Finally, glaciers present an intellectual challenge. How do they form? What keeps them from melting away? How do they move? Why are some retreating and others advancing? What was the glacier doing when it created a particular landscape? Few people can look at a glacier or walk in a glacially sculptured landscape for long without asking "why?" It is human nature to desire to understand the world around and beyond us.

There are many ways to enjoy glacier-watching. Visiting an area's glaciers whether on foot, by boat or plane gives one an overall view of the glaciers' and the regions' topography. Just as bird-watchers keep lists of sightings, so glacier-watchers have their lists of glaciers visited. This is an important activity, as the more glaciers one sees, the more one develops a feeling for their unique and special features. Establishing a photographic station by taking pictures of a glacier several times during a summer or over a period of years from the same spot helps one to see cyclic and long-term changes in both the glacier and the area around it. Many glaciers are seldom, if ever, visited by glaciologists; photographs of them can add to scientific knowledge about glaciers, particularly if the photographs are taken from established photographic stations. For this reason, the location of important established photographic stations is given. These are marked by cairns (rock piles). For the curious, there is the geological detective game of tracking down the origins of glacially sculptured landscapes. The reward — a fuller seeing and appreciation of the geological forces that gave us our splendid landscape: the peaks and ridges as well as our valleys replete with their mosaics of forests and peatland bogs, talus slopes and ponds. Finally, for the adventurous, glacier trekking and skiing provide an opportunity to visit remote corners of the earth, places where no one may have set foot and where few have seen the splendor of a black, pinnacled ridge soaring heavenward above a sea of shimmering white ice.

PHOTOGRAPHIC STATIONS: Since the turn of the century, glaciologists have kept a photographic record of the glaciers made from established photographic stations. Although many of the early photographic stations are no longer usable because the glaciers have retreated and alders have grown obstructing the view, new ones have usually been established. The most important photographic stations are located on the maps. In some cases, where a glacier is retreating or advancing rapidly, you may wish to establish your own station. To add your photographs to the continuing record, you may send copies to Nancy R. Lethcoe, P.O. Box 701, Whittier, AK 99693. Please include the location from which the photograph was taken and date. Photographs will be forwarded to the appropriate authority.

Fig-4. Nellie Juan Glacier, a tidewater (calving) glacier in Port Nellie Juan, has provided numerous spectacular ice-calving events when seracs (large pillars of ice) plunge from its face. Barren rock in the foreground and behind Nellie Juan has recently been deglaciated as the glacier continues to shrink and retreat. Like all tidewater glaciers in the sound, Nellie Juan rests on bedrock and is not floating. Harbor seals and sea otters often haul-out on the icebergs. Photograph (1981) courtesy of Nancy Simmerman.

Chapter 2. Glacial Formation and Movement

Glaciers develop when the yearly accumulation of snow exceeds the annual melt rate over a period of years, just as one's savings account grows when more money is accumulated than spent. Direct snowfall is the primary source of accumulation. Other sources include wind-blown snow, avalanches, hoarfrost (direct condensation of ice from water vapor), rime (the freezing of supercooled water droplets on striking a surface), hail, and the freezing of rain and melt water. In a glacier's AREA OF ACCUMULATION, more glacial mass and snowfall accumulates than melts. Just the opposite holds true in the glacier's ABLATION ZONE: the mass melted or calved yearly exceeds the amount accumulated. Melting (from sunlight, warm air, and rain) and calving (the breaking off of icebergs) account for most of the ablation, along with evaporation, ice avalanches, and the melting of the fronts of ice-calving glaciers by salt water, lakes or rivers.

Within the two zones, the amount of accumulation and ablation is not equally distributed. Topography influences the amount of accumulation in various parts of a glacier. For example, areas below an avalanche chute collect more snow than areas at the same altitude outside the avalanche run-out zone. Topography also influences melt-rate. Areas exposed to the sun melt faster than those shaded by a mountain. Thus, the part of Serpentine Glacier's terminus which is landbased and shaded by Mt. Muir is more advanced than its tidal terminus which is exposed to both sunlight and some tidal action (Fig-69).

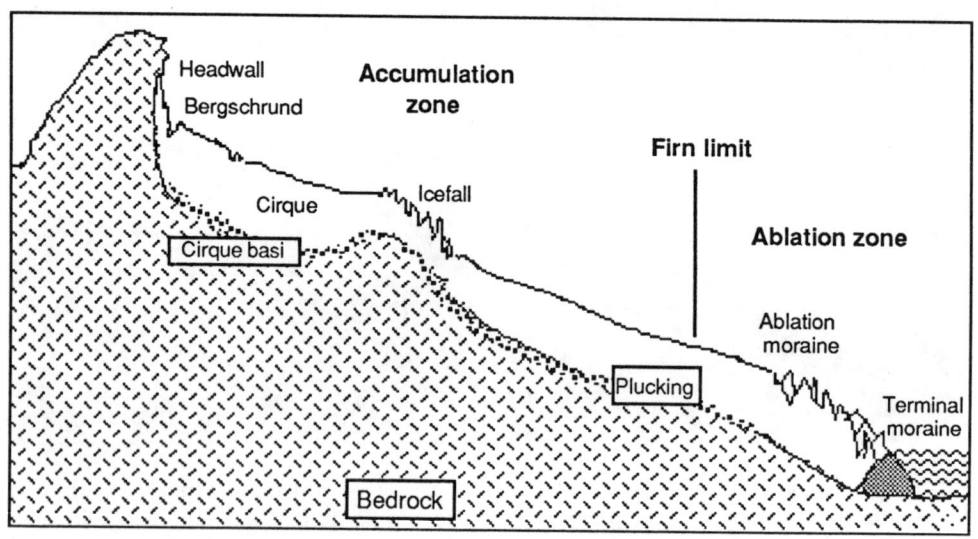

Fig-5. Profile of a typical Prince William Sound valley glacier flowing from a cirque-type accumulation area and terminating in ice cliffs that calve into the ocean.

SNOWLINE refers to the highest position reached each year by the retreating snow cover. Snowlines are usually dated by year. Thus, for example, one refers to "the 1983 snowline." The snowline on a glacier lies at a lower elevation than the snowline on adjacent hills because of the difference in melt rate between the reflective qualities (albedo) of the glacier and land. White glaciers reflect a large part of the sun's rays, while dark rocks and forests both reflect less light back and hence absorb more of it.

Snow that remains for more than one season is called FIRN. Firn is snow that has been partly consolidated by alternate thawing and freezing but has not yet attained the density of glacier ice. Firn particles have uniform dimensions and are usually less than one tenth of an inch in diameter. Successive snowfalls increase the pressure on the firn causing changes in density, volume, and crystalline structure.

FIRN LIMIT refers to the annual lower limit of the firn mantle. The firn limit separates the ablation area from the accumulation area (Fig-5). At times it may coincide with the snow line, but it may also differ. The firn limit varies yearly and locally depending on altitude, exposure, distance from the sea and other factors. By studying 1941 aerial photos of Chugach glaciers, Field found that the firn limit of glaciers on the coastal side of the central Chugach lay at about 1700 ft., whereas less than 38 mi. away on the northern side of the mountains, it was 3670 ft. — almost 2000 ft. higher (Field 1975). The firn limit extends lower on the coastal side of the Chugach Mountains because winter snowfalls are so much heavier here than on the drier, interior side.

Fig-6. The 1964 snowline on Nellie Juan Glacier is much lower than on the adjacent rocky peaks. Photo by Austin Post, U.S.G.S. August 25, 1964.

As the seasonal snowfall melts, annual firn lines appear on the glacier. Because thin layers of wind deposited dirt and dust separate each accumulation layer, annual firn lines are identifiable. Sometimes a volcano, such as Mt. Augustine in lower Cook Inlet, erupts and sends ash clouds over the Kenai/Chugach Mountains. Ash settles on the snow making a much darker than normal layer between its own and adjacent layers. Such ash layers remain visible for years in exposed bands on ice cliffs, seracs, and crevasse walls and can be useful in dating accumulation layers or in determining how many years it takes for firn to become glacier ice. In August, the accumulation layers are readily discernible on many of the smaller apron and cirque glaciers in Harriman and College Fiords.

Glaciers in the Prince William Sound area rapidly convert snow to ice. In the cold and dry climate of Antarctica, it takes hundreds of years for firn to become GLACIER ICE. By contrast, only a few years are required in Prince William Sound's warm, moist coastal mountains. The presence of rain and meltwater, which speed up the rate of densification of the snow particles, plus large differences in annual accumulation, resulting in large overburden pressure, account for this astounding difference in time.

The formation of glacier ice is not the same as the freezing of water in a lake or stream. The transformation of firn into glacier ice is a complex metamorphic process involving compression and repeated melting and recrystallization, which changes the size of the ice crystals and removes air passages. During recrystallization, particles change in shape, size, and density as water molecules transfer from one particle to another until at last the firn becomes glacier ice. Whereas firn has a density of 0.5 megagrams per cubic meter, glacier ice has a density of at least 0.84 megagrams per cubic meter.

Glacier ice is composed of large, interlocking single ICE CRYSTALS. Unmelted crystals in fresh icebergs and at the face of a glacier can be jiggled back and forth like pieces in a wooden puzzle and, if handled with care, can even be removed intact. The boundaries between ice crystals can be seen by pouring a little food coloring over a block of ice. The size of glacier ice crystals generally increases with age, distance travelled, and higher temperatures; whereas speed and stress decrease their size.

The surface of an ice crystal shows numerous tiny melt grooves and larger parallel grooves formed by melting along the basal planes of the ice crystal. Inside the ice crystal, air bubbles usually are arranged into bubble-rich and bubble-poor layers. These bubble layers are parallel to and part of the foliation structure of a glacier. Large ice crystals may contain several foliation bands.

Glacier ice is a crystalline solid, whose molecular bonds are so weak that it flows under its own weight. Under the stresses of glacial movement, it, like any other crystalline solid, exhibits both elastic and plastic deformation. Elastic or brittle deformation stretches the ice until it cracks. Crevasses and seracs result from elastic deformation, which occurs only in the glacier's upper 100 to 150 feet. Plastic deformation (creep) causes permanent changes in the ice by two processes: flow, where layers within the ice crystal glide

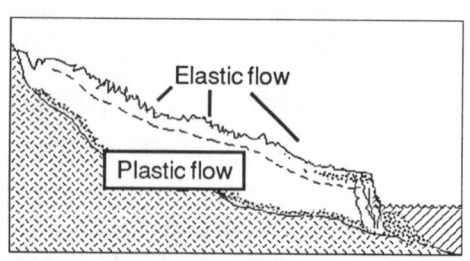

Fig. 7. Glacial ice is a crystalline solid that has both plastic and elastic flow.

over one another (like a deck of cards) and recrystallization, where the ice crystal's boundaries change in size and shape.

FOLIATION is a layered banding of ice often seen on exposed glacier surfaces or in icebergs. Foliation may begin with the deposition of dense and light snow or dirty and clean snow high in the accumulation zone. As the accumulation layers recrystallize under the pressure of plastic flow, foliation layers develop, much as metamorphic rocks show foliated layers whose origin lies in the original deposition of layers of sediments. Differences in the size and arrangement of the ice crystals plus variations in bubble content differentiate the bands.

One of the most frequently asked questions by visitors to the sound's glaciers is — why is the ice blue? Vivid blue colors occur when the ice is pure, as on a freshly exposed surface, because the ice crystals absorb the longer wave lengths and scatter back short blue and green wave lengths. Surface ice which has begun to melt or has impurities looks white — it reflects the entire spectrum of light.

Fig-8. Fluted iceberg and bottom berg showing fluted marks possibly caused by freshwater melt streams rising from the base of Columbia Glacier. Photo by author.

Icebergs:

Tidewater glaciers calve (discharge) millions of tons of ice each year into Prince William Sound. Besides their awesome beauty and imaginative shapes, some icebergs exhibit unique features. FLUTED ICEBERGS may come from the face of the glacier where fresh meltwater streams, which are lighter than salt water, flow up the glacier's front carving away the ice in distinctive, curtained patterns. However, proof for such a theory is lacking, since so far now one has been foolhardy enough to don a wet suit and examine the underwater surface of a tidewater glacier. BOTTOM BERGS are dark, dirty looking bergs from the glacier's sole. Ice captured here is very old having been deposited near the glacier's head. During the course of a bottom berg's long journey to the terminus, air bubbles have been compressed out of it making the ice a darker blue, and it has plucked rock debris from the valley floor, making the ice look almost black. If an iceberg carries a heavy burden of rocks, their additional weight may make it float lower in the water. DARK BLUE BERGS come from near the bottom, where most air bubbles have been displaced from the ice. BRECCIA BERGS, known locally as conglomerate bergs, appear as a mishmash of jumbled ice fragments refrozen together. Recently, they have become a common type of iceberg sighted at Columbia Glacier; they can also sometimes be seen near Meares and Barry glaciers.

The word "iceberg" is used generally to refer to all floating ice originating from a glacier or specifically, to only very large icebergs, as in the following classification. Unless otherwise indicated, the word "iceberg" is used in its wider sense in this book.

TYPE	APPROXIMATE SIZE
Brash ice	Accumulations of floating ice composed of fragments less than 6 ft. across.
Growler	Smaller pieces of ice than bergy bits, appearing greenish in color, larger than 6 ft. across but less than 3 ft. above water.
Bergy bits	A large piece of floating glacier ice, generally showing between 3 and 15 ft. above sea-level. About the size of a small house.
Iceberg	Large piece of glacier ice, protruding more that 15 ft. above sea level.

Fig. 9. Adapted from LaBelle (1983).

Most floating ice from glaciers calving into Prince William Sound is of the brash and growler type. Occasionally, a retreating glacier releases a large bottom berg that may be a bergy bit or iceberg. Very few bergy bits from Columbia Glacier enter the shipping lanes, because shallow water over the moraine usually confines them between the glacier and the moraine until they break into smaller growlers and brash.

The life-expectancy of an iceberg is short. How long it takes to melt depends on the water temperature. During the winter, when surface water temperatures drop to near freezing a large piece may take several days to melt. In the summer, when surface water temperatures rise to over 40 degrees, the same iceberg would melt much faster. Most ice in Prince William Sound melts within a few hours; only a very few last a day or more.

When icebergs melt, they often take on a saddle or head-and-tail appearance reminiscent of a swan, because melting occurs faster underwater. Thus first one end of the iceberg melts, then rises, while the other end sinks and begins melting. As melting progresses, the iceberg alternates back and forth on its major axis, which leaves the central portion or minor axis constantly exposed to erosion. A two-humped iceberg results.

During the formation of glacier ice, trapped air is compressed into bubbles. As icebergs melt, the pressurized bubbles crackle and pop.

How much of an iceberg is visible above water depends on its specific gravity and shape. Because icebergs have a specific gravity of 0.9, nine-tenths of the *mass* of all icebergs is submerged. However, the shape of an iceberg also influences the relationship between its exposed and underwater *areas*. Flat, table-top-like icebergs and blocky, rectangular icebergs have six parts (area, not mass) submerged for every part exposed. Rounded icebergs have only four parts below water; and winged icebergs, which have sail-like pinnacles around their central areas, have almost a one to one ratio. In addition, the amount of rock debris carried by an iceberg affects its specific gravity. A black bottomberg, laden with morainal material, will float much lower in the water and may be nearly impossible to see under adverse conditions.

Iceberg movement depends on the size and shape of the berg as well as on currents and wind. Once in motion, the momentum of an iceberg is so strong that it can continue to move for sometime after the wind or currents have ceased. Currents, even weak ones, usually predominate over variable winds of less than 32 knots. If the wind persists above 32 knots for over twelve hours, then the wind becomes the dominant influence even if the currents are two to three knots. Thus, predicting the movements of icebergs is a challenging enterprise. The discharge of ice from Columbia Glacier into the tanker lanes of Valdez Arm have prompted local studies of the complex factors involved in predicting the movements of icebergs.

Radar is often used to detect large icebergs, but its usefulness for small icebergs, the size of a growler, is limited to flat seas and ranges of one mile or less. When winds build seas to four feet or more, the echoes from the waves often conceal the echoes from an iceberg. Echoes from rain, snow and fog can also hide the presence of smaller bergs. Under these conditions, icebergs in the North Atlantic as large as 22 ft. high have at times remained undetected. For this reason, when the Coast Guard station in Valdez issues iceberg hazard warnings, it routinely reminds mariners that icebergs are not always detectable by radar.

Glacial Movement:

Glaciers move downhill under the force of gravity. Even when a glacier is retreating or stable, new ice continues to move downhill replacing the old ice as it melts.

The movement of a glacier is complex and varies according to the climate, thickness of the ice, bedrock conditions, channel shape and slope, and conditions at its terminus. Glaciers respond to changing conditions in their accumulation areas and at their termini by adjusting their rate of flow, by changing their thickness in the accumulation and ablation zones, or by advancing and retreating at the terminus.

Glaciers, even retreating glaciers, are constantly flowing downhill under the force of gravity. How fast a glacier flows depends in part on the magnitude of its accumulation and ablation and their distribution over the glacier. The sound's coastal glaciers receive much more snowfall each year than do interior glaciers; to remain stable, they must move much more snow from their accumulation areas to their ablation zones than colder, interior glaciers. This is why so many glaciers in Prince William Sound extend all the way from the high peaks to the sound's warm, sea level waters. The rate at which ice flows in this coastal region is significantly higher than for interior glaciers. This accounts for the frequent calving displays enjoyed by the sound's glacier-watchers.

A glacier's rate of flow does not remain constant, but is influenced by short-term seasonal changes, longer-term climatic fluctuations, and conditions at its bedrock. During the winter and spring, a glacier moves more slowly than in late summer when surface melting, heavy rains, and meltwater generated by the heat produced by basal sliding all combine to lubricate its sole increasing its rate of flow. Thus, August and September are often the best months to watch tidewater glaciers calving. Over the years, the amount of snowfall a glacier receives fluctuates depending upon other climatic conditions, such as the position of the Japanese Current and sunspot activity. Excessively high accumulations of snow generate a larger bulge of ice called a KINEMATIC WAVE that sometimes moves towards the terminus faster than the glacier's normal rate of flow, leaving behind a trail of severely cracked ice. Sometimes the terminus may advance rapidly for a season or two then slowly retreat as the glacier's velocity returns to normal. Sporadic sudden advances in Childs and Miles Glaciers are thought to be the result of kinematic waves. GLACIER SURGES are thought to occur when the basal meltwater stream for some as yet unknown reason leaves its channel and spreads out in a thin layer between the glacier's sole and the constraining bedrock, decoupling the glacier from its bedrock. The resulting loss of basal friction causes the glacier to slide much faster than normal until the stream returns to its channel and contact with bedrock is reestablished. Surges occur at fairly regular intervals (Kamp 1985). It is possible that Childs Glacier might be a surging glacier but more study is needed (Field 1975).

Topography influences the rate of flow of a glacier in several ways. A glacier accelerates when winding down a steep valley then slows when it spreads out onto an outwash plain, just as a river rushes through a gorge then meanders across a flat lowland. Ice at the center of a glacier flows faster, generally, than ice which is held back by the friction of the valley sidewalls. Likewise, the surface speed of a glacier increases as it flows over a convex surface, such as over a subglacial cliff, forming an icefall, and slows as it settles into the concave surface at the icefall's base. Again, this is similar to water hurtling over a falls then settling into a plunge pool at its basin.

Finally, conditions at a glacier's terminus affect its rate of flow. A retreating glacier terminating on land, thins and slows towards its front. Frequently, the front of a valley glacier forms a bulb-shaped terminus as it spreads out over a delta or outwash plain. Tebenkof Glacier is one of the best and most frequently seen examples of a valley glacier terminating on land. By contrast, valley glaciers that terminate at tidewater generally do not thin and slow, because the warm sea water melts and undercuts the face causing dramatic calving events and rapid melting.

The Importance of Aspect and AAR ratios:

The ASPECT of a glacier refers to its direction of flow. A glacier with a southern aspect faces south. Aspect is particularly important for small glaciers. Apron and cirque glaciers on a north-aspect slope receive a smaller total amount of heat than south sloping glaciers; hence, the ablation rate of glaciers with a northerly aspect is less than that of southern facing glaciers. Observers in Harriman Fiord can readily see the importance of aspect and location on the firn limit, which is lowest on glaciers at the southern end of the fiord and higher on glaciers at the northern end. The more southerly glaciers lie closer to the Gulf of Alaska, the source of incoming precipitation, and have a northwesterly aspect, whereas Barry Glacier, the most northerly one, faces southwesterly.

COMPARISON OF ASPECT AND FIRN LIMITS FOR HARRIMAN FIORD'S MAJOR GLACIERS			
GLACIER	GLACIER ASPECT	1941 FIRN LIMIT	1970 FIRN LIMIT
Barry Gl.	SW	2460 ft.	3500 ft.
Surprise Gl.	NW	1300 ft.	2650 ft.
Harriman Gl.	NW	980 ft.	1700 ft.

Fig-10. As the Earth's climate has warmed, the firn limit has risen significantly during the mid-part of this century. A higher firn limit means relatively more of the glacier now lies in the ablation zone and less in the accumulation area. As a result, all the land-terminating glaciers in the region are now retreating. (Based on W.O. Field, 1975 and unpublished data provided by courtesy of Austin Post, 1986).

For larger glaciers, the distribution of snow with respect to altitude is more important than aspect. Glaciologists use the ratio between the accumulation area and its total area (ACCUMULATION AREA RATIO = AAR) as an indication of whether a glacier is likely to grow or recede. If a glacier has a very high percentage of its area in the accumulation zone, such as Chenega (94%), it is usually stable or advancing.

For tidewater glaciers, the position of the terminus must be considered along with the AAR. Tidewater glaciers that have undergone major retreats in the last couple of centuries can remain stable or advance with lower AARs. Thus, Harriman and Harvard are now advancing at about one mile a century — not so much as a result of their accumulation area ratios (AAR) but because they previously retreated to stable, retracted positions. However, if the AAR is low, a glacier may have difficulty sustaining its position or advancing. Harriman's AAR is so low that it has difficulty advancing and sometimes forms embayments (large bays in the tidal ice cliff). Extended tidewater glaciers, such as Columbia, need a high AAR to maintain their advanced position. Columbia's low AAR rating (66%) raised the early suspicions of Austin Post and other glaciologists that the glacier might not be able to maintain its extended position.

ACCUMULATION AREA RATIOS (AARS) FOR SELECTED TIDEWATER GLACIERS

Tidewater Glacier	Area in sq. miles	Firn Limit (late '60s)	AAR (late '60s)	Glacier's status
Barry	29.0	3500	74%	retreating
Cascade	6.5	3900	88%	slow advance
Chenega	143.0	1900	94%	stable
Columbia	435.0	2800	67%	retreating
Harriman	19.7	1700	68%	advancing
Harvard	199.0	3600	80%	advancing
Meares	56.0	3300	85%	advancing
Nellie Juan	22.0	1300	91%	retreating
Surprise	32.0	2650	83%	slow advance
Tiger	21.6	1900	89%	stable
Yale	74.0	3700	81%	retreating

Fig-11. Courtesy of Austin Post, 1986, unpublished research.

Since a glacier's firn limit fluctuates with short-term changes in the climate, the way the accumulation zone is distributed with respect to ALTITUDE is also important. On many of Prince William Sound's glaciers, such as Meares and those along the west side of College Fiord, most of the accumulation area lies above 4,000 feet. Small fluctuations in temperature have little effect on these glaciers' total accumulation. However, glaciers with major accumulation zones below about 3000 ft. are susceptible to temperature changes of just a few degrees. Dirty, Wedge and Toboggan glaciers in Harriman Fiord have low altitude accumulation areas and have been retreating throughout this century. This is also true of Nellie Juan Glacier.

Chapter 3. Types of Glaciers

Glaciologists divide glaciers into two main categories: polar (cold) and temperate. Polar glaciers are very cold and are frozen to their bedrock. Most polar type glaciers are located in Greenland and Antarctica. Temperate glaciers have an internal temperature at or very near the melting point — except for a colder, shallow, surface layer during winter. They slide over their bedrock.

Two basic types of temperate glaciers occur in Prince William Sound: small glaciers with local accumulation zones and larger glaciers whose accumulation areas lie within an icefield. The sound is a popular place for glacier observation, because it has some of North America's finest examples of different types of temperate glaciers.

Numerous small glaciers isolated from any supporting icefields drape the sides of the Chugach and Kenai Mountains. Direct snowfall, plus avalanches and/or wind-blown drift account for most accumulation. Small, thin, wide glaciers, clinging to a mountain's sides are called APRON GLACIERS. Apron glaciers occur throughout the sound, and most have not attracted the attention of glaciologists. A casual survey of these glaciers with binoculars from a boat shows that almost all of those at or below 4500 ft. are surrounded by barren rock which indicates they are retreating. Those at higher elevations seem to be more stable.

CIRQUE GLACIERS also have local accumulation zones. These glaciers may form the head of a valley glacier or may be confined to a cirque basin (Fig-5). Above cirque glaciers tower sheer headwalls. Most frequently these are awesome sharp, jagged ridges called "arêtes." The upper headwall is shaped by frost-thaw action. Frost-thaw action occurs whenever rain or meltwater seeps into cracks in the rock headwall. When the temperature drops, the water freezes. Its expansion into ice widens and deepens the crack weakening the rock, just as water frozen in a glass jar breaks the jar. When the ice melts, the loosened rock tumbles to the glacier below. Meanwhile, the cirque glacier erodes its basin in a rotational manner. Because the flow above the equilibrium line is downhill, while the flow below it is upward, a rock basin is excavated.

ICEFIELDS are large, nearly level areas of ice, where the underlying topography allows ice to accumulate much like water filling up a lake. Three large icefields and one smaller one feed Prince William Sound's glaciers. The Sergeant Icefield in the Kenai Mountains which abut the western part of the sound covers roughly 825 sq. miles and has the lowest elevation accumulation area in Alaska. The much smaller Whittier (Field: Blackstone/Beloit) Icefield serves as an ice resevoir for glaciers on both sides of the Kenai Mts. Stretching across the northern part of the sound from College Fiord to Columbia Bay, the Chugach Icefield (686 sq. miles) feeds glaciers flowing from the Chugach

An Observer's Guide to

Fig-12. In the Kenai Mountains on the westernside of the sound, the Sargent Icefield serves as a huge resevoir feeding (from right to left) Contact, Nellie Juan, an unnamed glacier, Ultramarine, Princeton, Chenega (not visible), and Tigertail glaciers. The Gulf of Alaska is visible in the background. Photo by Post USGS, 1961.

Mountains. It differs from the other large icefields in being composed of a series of interconnected smaller icefields. To the east, the Bagley Icefield feeds glaciers flowing into the Copper River and Gulf of Alaska. Its Bering Glacier, the world's largest glacier outside of Greenland and Antarctica, flows from the Bagley Icefield into the Gulf of Alaska about 80 miles east of the sound. The Bering Glacier alone covers approximately 2,308 square miles compared with the State of Rhode Island which has only 1,214 square miles.

Icefields, like cirques, may have valley glaciers as their outlets. Valley glaciers flowing from icefields are longer than they are wide and frequently have tributary glaciers. Above the glacier, the walls of the valley are usually very steep rock or scree slopes. Intensive frost-thaw action weakens the rock walls causing rock slides which leave a dark band of rocks deposited along the glacier's edge — called a "moraine."

Valley glaciers terminate on land, in a lake, river or ocean. Tebenkof is an example of a valley glacier terminating on land. Valley glaciers that terminate in water are called CALVING glaciers. Portage, Amherst, Valdez and Miles glaciers calve into lakes. Childs Glacier terminates on the Copper River, while Columbia, Harvard and Chenega have ocean or tidal ice fronts.

CALVING (TIDEWATER) GLACIERS: Early European explorers were much perplexed by the loud, booming noises they heard in the vicinity of glaciers. Hearing thunderous booms in Columbia Bay, the Spanish explorer Salvador Fidalgo postulated a volcano at this location. Capt. George Vancouver recounts in his journal the final solution to the mystery: "Whilst at dinner in this situation [College Fiord] they [Whidbey's exploratory party] frequently heard a very loud rumbling noise, not unlike loud, but distant thunder; similar sounds had often been heard when the party was in the neighborhood of large bodies of ice, but they had not been able to trace the cause. They now found the noise to originate from immense ponderous fragments of ice, breaking off from the higher parts of the main body, and falling from a very considerable height, which in one instance produced so violent a shock, that it was sensibly felt by the whole party, although the ground on which they were was at least 2 leagues (6 miles) from the spot where the fall of ice had taken place." (Vancouver 1798, pp. 183-184.)

It is interesting to note that Vancouver was apparently unfamiliar with the word "glacier" although LaPerouse had used it earlier on his maps. Neither Cook nor Vancouver show any glaciers on their maps of Prince William Sound.

As time passed, the early explorers seeking the northwest passage were followed by men of science who came to map and study the sound. Mendenhall, Gilbert and Gannett, Grant and Higgins, and Tarr and Martin photographed the glaciers. And, as they collected information, a second puzzle emerged. Some of these scientists noticed that some tidewater glaciers were retreating while others flowing from the same icefield were advancing. They reasoned that if climate alone determined whether glaciers were retreating or advancing, then all the glaciers in a particular area ought to advance and retreat together. The behavior of tidewater glaciers led glaciologists to suspect that terrain, altitude, firn limit, and a glacier's accumulation area ratio are also significant factors. But even these did not provide a complete explanation of the seemingly erratic behavior of tidewater glaciers flowing from the same icefield.

Then, in the 1970s, Austin Post (US Geological Survey Project Office of Glaciology in Tacoma, Washington, under the direction of Dr. Mark Meier) developed a new theory taking into account the effect of salt water on a glacier's terminus. According to Post, the depth of a tidewater glacier at its terminus and the condition of its accumulation zone together may determine whether a glacier is retreating, stable, or advancing more than individual climatic changes. Glaciers ending in shallow water or on moraines may still be retreating (Nellie Juan and Shoup) or can be slowly advancing (Harriman, Harvard, and Meares.) Before a tidewater glacier can shift from retreating to advancing, it

An Observer's Guide to

The Cyclic Behavior of Tidewater Glaciers

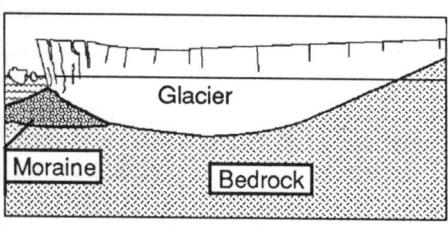

a. A calving glacier in its extended position pushing its terminal moraine down the fiord. Most of the ice face is protected from the salt water by the moraine.

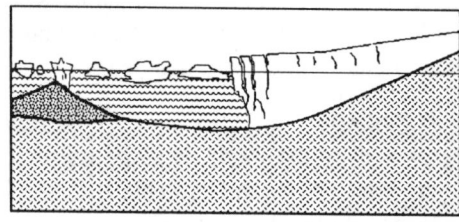

b. A calving glacier in irreversible retreat. Note how much more of the terminus is exposed to the melting effect of the salt water. Large bergs ground on moraine.

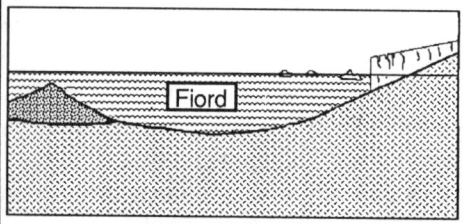

c. The glacier retreats to shallow water or on to land until it regains a balance between its accumulation and ablation areas.

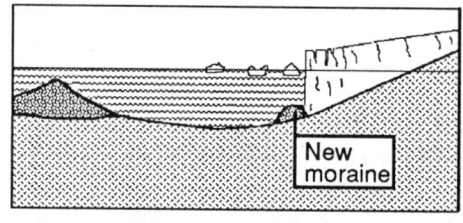

d. Rock debris carried to the glacier's terminus begins to build a new moraine which protects the ice front. Once again, the glacier begins to advance.

Fig-13. Illus. by R. J. Lethcoe

must first reach a stable position where its accumulation and ablation zones are in balance. The glacier then begins to build a terminal moraine through eroding its valley walls and bedrock cradle and depositing the rocks at its terminus. As this moraine grows, it protects more and more of the glacier's ice front from the rapid melting effect of the warm salt water. Gradually, less and less ice melts each year in the ablation zone, while ice from the accumulation area continues to push seaward. Slowly, at times no more than a few inches a year, ice descending from the accumulation area pushes the terminal moraine forward. Eventually, the advancing glacier reaches an extended position of equilibrium between accumulation and melting. Like Columbia, it may rest here nearly stable for many centuries.

Tidewater glaciers remain stable until the climate changes raising the firn limit, the glacier advances so far that it over-extends its "ice-budget," or the glacier pushes its protecting terminal moraine over a submarine canyon. In all cases, the ablation area increases, so the accumulation area ratio lowers. With increased melting, calving reaches a critical point. As the terminus thins, the glacier's slope increases — further accelerating its rate of flow. More and more ice is drawn down from the icefield's reservoir, augmenting the thinning process there. When the glacier's snout thins sufficiently, it retreats off the supporting moraine. Drastic retreat begins.

Tidewater glaciers retreat relatively fast in depths of more than 240 ft. (80 m.), because relatively more of its terminus is exposed to the water which serves as an inexhaustible source of heat. When the channel narrows or turns, the glacier retreats more slowly, probably because less of its terminus is exposed, and discharge lessens, thus reducing the iceberg-calving rate. Retreat remains slow until the terminus reaches another broad or straight part of its channel whereupon the glacier will again retreat drastically until it reaches a stable position in shallow water. Once the glacier stabilizes, it slowly begins to build a new moraine. As the moraine grows, it provides the glacier's terminus with increased protection from warm water. The ablation rate slows. The glacier is ready once again to begin its slow push down the fiord. The cycle of slow advance and drastic retreat takes place over centuries for valley glaciers with a low angle of slope, such as Yale and Harvard, and over decades for smaller, glaciers descending precipitous slopes, such as Bryn Mawr and Wellesley.

Some tidewater glaciers in the sound are now in their stable retracted positions. Others are advancing from retreated positions. In the mid-'70s Austin Post pointed out that Columbia Glacier held the unique position of being the only Prince William Sound tidewater glacier to be in a maximum extended position (Post 1975). After several years of study and modelling of iceberg-calving dynamics, the USGS predicted that rapid retreat would begin in 1982 (Meier 1980).

Fig-14. In 1968, Columbia Gl. was still advancing on to Heather Island. A. Post. USGS.

Chapter 4. Glacial Features:

The continuous downhill movement of a glacier stretches its brittle surface layer beyond the breaking point creating elongated cracks in the glacier's surface called CREVASSES. Crevasses only occur in the surface area and penetrate to the point where the rate of plastic flow tends to close the crevasse. Most crevasses in coastal glaciers are less than 160 feet deep. Crevasses concealed by a snow bridge can be dangerous.

Glaciologists distinguish several types of crevasses. The BERGSCHRUND is a deep crevasse at the head of a glacier where moving glacial ice pulls away from the stagnant 'headwater' ice (Fig-5). Observers aboard a boat who have a good pair of binoculars can often see a series of smaller cracks above the bergschrund. These are the future bergschrunds preparing to replace the current one as it flows down glacier and closes. TRANSVERSE CREVASSES, which are characteristic of extending flow in the accumulation zone, stretch across a glacier and generally occur where the slope of a glacier steepens. Often they are concave, but they may become convex if the glacier moves rapidly, such as above an icefall (Figs-15,16).

SHEAR CREVASSES, common features in the ablation zone's compressive flow, develop when a rapid change in velocity occurs at a glacier's edge (Figs-15, 16). As a glacier spreads out into a wider area, LONGITUDINAL CREVASSES form (Fig-19). Longitudinal crevasses are common near the terminus of valley glaciers, such as at the terminal faces of Tebenkof, Columbia and Sheridan glaciers. Large isolated blocks of ice called SERACS form where two crevasse patterns cross (Fig-4). Glaciologists study crevasse

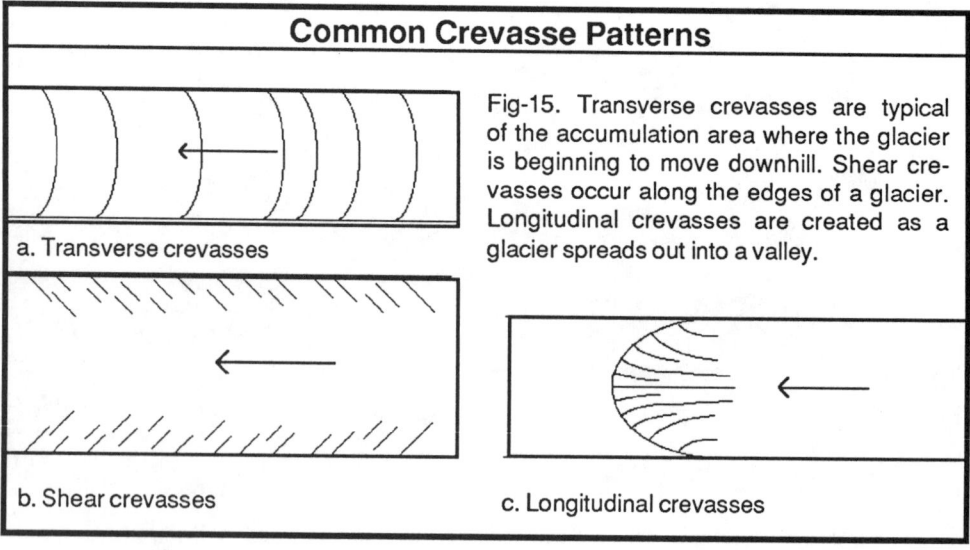

Common Crevasse Patterns

a. Transverse crevasses

b. Shear crevasses

c. Longitudinal crevasses

Fig-15. Transverse crevasses are typical of the accumulation area where the glacier is beginning to move downhill. Shear crevasses occur along the edges of a glacier. Longitudinal crevasses are created as a glacier spreads out into a valley.

Fig-16. Normally concave transverse crevasses become convex as the glacier accelerates towards an icefall. Peaks and ridges surrounded by a glacier are called "nunataks." Chugach Icefield, Columbia Glacier. A. Post, USGS. Aug. 9, 1973.

patterns to obtain information concerning a glacier's rate of flow. Changes in Nellie Juan Glacier's crevasse patterns can be seen by comparing Figs-50-53.

ICEFALLS occur where the glacier descends rapidly down a steep slope. They are heavily crevassed because the steep slope and confined channel accelerates the basal ice faster than the surface ice can respond. As a result, the surface ice fractures into a plethora of crevasses and seracs that are pulled apart by the extended flow. Boaters drifting in front of Cascade or Coxe glaciers can often hear and sometimes witness seracs falling into the crevasses. Beneath the icefalls, the glacier slows. Compressive flow quickly closes the crevasses; the glacial surface assumes a smooth, bowl-of-porridge appearance (Fig-18). Many of the sound's glaciers exhibit a stairstepping effect of alternating small bowls and icefalls. Wellesley, Vassar, and Cascade glaciers have such typical stairstep features.

ICE AVALANCHES occur inside icefalls and at the terminus of a glacier. Ice falling from a tidewater glaciers' terminus is called CALVING. In the Prince William

Fig-17. A large ice avalanche from Detached Glacier (Surprise Arm, Harriman Fiord) triggered a rock avalanche that lasted over an hour and significantly changed the course of the stream flowing from the glacier. Photos by Nancy Simmerman.

Sound area, most glaciers which terminate on steep slopes have relatively frequent ice avalanches. Ice avalanches have been particularly common at Bettles, Cataract and Detached glaciers in recent years. The Harriman Alaska Expedition gave Roaring Glacier its name because of the large number of ice avalanches that occurred there during their visit. Subsequently, Roaring Glacier has retreated and no longer puts on such displays. In 1984, we observed an ice avalanche from Detached Glacier which broke through the terminal moraine and started a rock avalanche. The ice and rock avalanche lasted for over an hour. At the end, one of the three waterfalls flowing from the glacier ceased and a new one began approximately a hundred yards away at the site of the avalanche.

Water enters a glacier from its surface or sides and may eventually find its way into streams. Streams flow along the top, through (englacial) and at the bottom of a glacier (Fig-29). Streams flowing on a higher level of the glacier descend to a lower level through holes in the ice called MOULINS. Basal water appears to flow in one or only a very few major tunnels (Kamp 1985). Ice caves and some embayments at the termini of tidewater glaciers mark the mouths of these streams.

Moraines:

As glaciers move, they accumulate rocks along their sides and on the surface or pluck

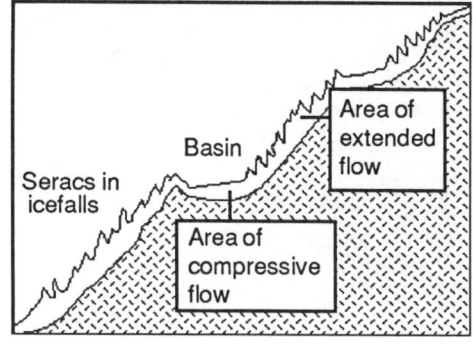

Fig-18. Many glaciers in Prince William Sound descend the steep mountain slopes in stairsteps of alternating icefalls and basins. When they retreat, valleys with a stairstep profile remain. Illus. R. J. Lethcoe

them from their bedrock basins. The term LATERAL MORAINE refers to rock debris found along the margins of a glacier, which has been scraped from the valley walls or been deposited by rock and dirt slides (Fig-21). In Barry Arm, one can see that the lower part of Cascade Glacier's lateral moraine derives primarily from glacial plucking and prying, whereas on adjacent Barry Glacier numerous snow avalanches and scree slopes also contribute rock debris to the lateral moraine. Thus, the topography of a glacier's valley walls can influence development of its lateral moraines.

When a glacier retreats, its lateral moraines are left stranded on the hillside. Excellent examples of older lateral moraines, now covered by vegetation, can be seen near Serpentine (Fig-68-69), Holyoke, Yale, Shoup, Worthington, and Childs glaciers. In a few areas of the sound, the tree limit appears to coincide with old lateral moraines, which provide the developing soils with crushed rock. Rock debris releases minerals into the soil and improves drainage. Above the lateral moraine, the mountain slopes are steep, their rock surfaces worn smooth by glacial erosion and weathering. Minerals trapped in

Fig-19. Longitudinal crevasses form as a glacier flows out of its constricted valley onto a wider plain. The expanding lateral motion of the glacier creates the longitudinal crevvases. Dark ice along the bottom of Tebenkof glacier's forward edge is bottom ice and ablation moraine. Photo by Nancy Simmerman.

Fig-20. Occasionally, hikers traversing a glacier encounter cone-shaped formations of ice that are covered or surrounded by rock debris. Large rock avalanches have covered and protected the glacier from melting. Gradually, the glacier melts down around the avalanche pile, and the rocks slide off revealing an ice-cone. Similar ice-cones occur on Falling Glacier, Kings Bay. Note how heavy ablation moraine covers the lower portion of Valdez Glacier. Photo by Marjorie Slatterly, 1985.

the bedrock are not so readily available for soil development and utilization by plants. Hence, only a few hardy lichens, mosses, shrubs and herbaceous plants grow here.

Most MEDIAL MORAINES occur when lateral moraines of tributary and main glaciers flow together forming a dark Y-shaped pattern on the glacier. Medial moraines clearly illustrate how each glacial stream, unlike merging rivers, retains its separate identity. Barry, Cascade, Surprise, Harvard, Meares and Valdez glaciers all have prominent medial moraines formed in this way. Many glaciers have medial-like moraines that surface in the middle of the glacier, usually somewhere in the ablation zone. These are called EMERGENT MORAINES. Some probably originate from glacial plucking at a submerged bedrock ridge or knoll. Baker and Amherst glaciers have excellent examples of this type of medial moraine. Others, such as the emergent moraine on Valdez Glacier, are

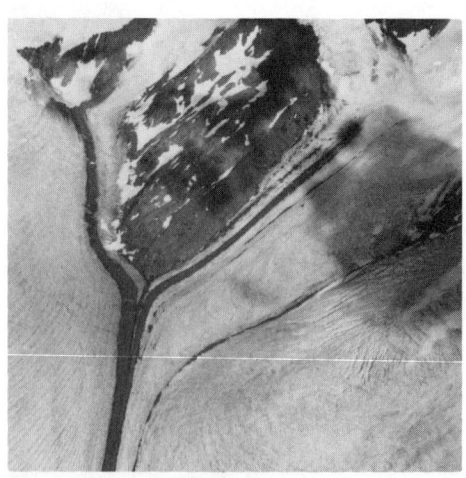

Fig. 21. *Top left:* Lateral moraines from glaciers flowing down either side of the ridge join to form a medial moraine. The position of the tributary glacier's moraine shows the glacier is very weak and does not contribute much ice to Valdez Glacier.
Center: Three more lateral and medial moraines flow into Valdez Glacier contributing varying amounts of ice. The position of the moraines shows that the ice streams do not merge. Glaciologists study pictures like this to compare the relative strengths of tributary glaciers over a period of time. Austin Post, USGS. Aug. 24, 1964.

probably medial moraines originating in the accumulation zone that have been concealed until the increased melting of the ablation zone removes the surface covering.

Some tributary glaciers lack the strength and volume to assume an inset position in the main stream (Fig-21). These weaker glaciers ride on top of the larger glacier stream forming a U-shaped morainal curve on its surface. They usually melt before reaching the glacier's terminus (Fig-69). At times a strong glacier may flow into the valley of a tributary glacier or a side valley where an ice-dammed lake may form (Figs-36, 107). This is best spotted from the air.

TERMINAL MORAINES, crescent-shaped rock piles found in front of glaciers or left by former glaciers, are created by rock material carried down from up-glacier. They consist both of material deposited on the surface of the glacier and material picked up and pushed forward by the glacier sliding over its bed. When a glacier stagnates for centuries or pushes slowly forward for thousands of years, the terminal moraines may build to hundreds of feet thick. Columbia Glacier's recently abandoned terminal moraine rises to the surface from depths of nearly a thousand feet. Submerged terminal moraines (Fig-23) left by Prince William Sound's neoglacial tidewater glaciers cross many fiords in the sound. They appear on the charts as fan shaped shoals with radically deeper water on each

Fig-22. Meares Glacier, which has been advancing during this century, shoves a small push moraine before it as it approaches Ranney Creek. Note the frost-kill zone near the glacier and the trees knocked over. Meares Glacier, 1984. Photo by Nancy Simmerman.

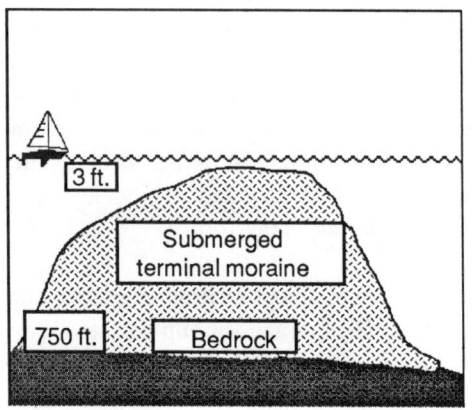

Fig-23. Former terminal moraines cross the mouth of Barry Arm, Harriman Fiord, College Fiord and Icy Bay. Mid-fiord moraines occur in Columbia Bay (new), Heather Bay (new) and Unakwik Inlet. Because they rise so abruptly from the bottom and cannot be seen from the surface, many a mariner has gone aground on them. Illus. R.J. Lethcoe

side. Examples can be found at the mouth of Icy Bay, between Granite Point and Pakenham Point at the mouth of Barry Arm, and across the middle of Unakwik Inlet. Columbia Glacier's moraine which stretches across Columbia Bay to Heather Island is the newest addition to the list. In August of 1987, The state ferry *E.L. Bartlett* went aground on it. Mariners are urged to exercise caution near these submerged moraines.

ABLATION MORAINES are both exposed and deposited as a glacier recedes. From a distance a receding valley glacier's tongue looks grey to black. Tebenkof, Princeton, Dirty, Surprise, Vassar, Shoup, Valdez, Sheridan and the eastern side of Columbia Glacier are all good examples. Closer inspection reveals that a thin layer of rocks and debris cover the glacial surface. Gradual wasting down of the surface ice concentrates the material into an ablation moraine. Ice between the dark, surface layer of rock and the basal ice is usually very clean. This can be observed in ice caves near the front of a glacier.

PUSH MORAINES, unlike ablation moraines, are made by advancing glaciers plowing morainal material before them. Old push moraines from Columbia Glacier's advances during the past century can be seen on the north end of Heather Island and along the eastern edge. Presently, Meares Glacier is advancing and shoving a small push moraine in front of it along its northern side.

Not all push moraines are made by the movement of the glacier's leading edge; some are built up by glacial ice riding up on a shoreline. In the late 1980s, this was occurring on the north end of Heather Island where ice bergs from Columbia Glacier created small pushlike moraines along the shore.

MORAINAL ROCKS vary in size and have edges ranging from smooth to sharp. Most surface rocks are large, because water flowing through the moraine washes away smaller particles. By looking at the edges of a rock, one can make some educated guesses as to its history. Rocks with sharp, angular edges come from glacial plucking or rock falls. Smooth, rounded rocks are stream-worn. In a few cases, streams flowing from ancient glaciers probably originally rounded them; subsequently, a new glacial advance picked them up and continued the process. In other cases, rounding occurred in englacial

Fig-24. Rocks exposed and superimposed on each other in Columbia Glacier's ablation moraine along the eastern margin. The large, sharp-edged rock at the top has recently been deposited on the glacier. The dark spot in the right-hand corner is moss. The rounded rocks have been carried down-glacier from high in the Chugach Mountains and have been smoothed and worn by streams flowing inside the glacier. Photo by author.

streams where the rocks jostled and tossed against other rocks (Fig-24). Still others lost their sharp edges in the silty, turbulent waters of outwash streams.

Frost-thaw action combined with weakening of the head and valley walls by the undercutting activity of the glacier, sometimes causes large rock avalanches (Fig-64,65). These thunder off the mountainsides on to the glacier's surface. If rock avalanches occur high up on the glacier, they become buried in the ice to emerge, years later, at the glacier's terminus. A black debris band possibly from a rock avalanche, has been visible for years on the east side of Columbia Glacier (Fig-8).

In 1964, the largest earthquake ever recorded in North America had its epicenter in Prince William Sound. Although the earthquake deformed land within a 200 mile radius of its epicenter, it generally had very little impact on the glaciers. Sherman Glacier, near Cordova, is the major exception. Here, the 1964 Alaskan Earthquake triggered massive rock and snow avalanches. The debris covered 3.3 square miles and was 16 feet deep. It is still visible 23 years later, giving the glacier the appearance of a chocolate sunday. In fact, local residents have dubbed Sherman Glacier the "Chocolate Sunday Glacier."

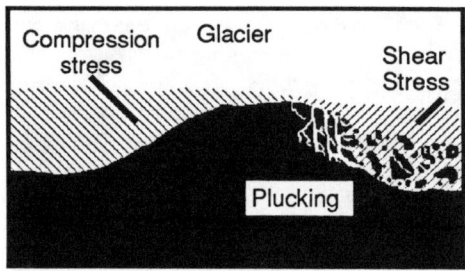

Fig-25. The upstream side of a knoll is smoothed by glacial action, while rocks and boulders on the downstream side are incorporated into the glacier through the freezing of meltwater in pre-existing crevices and plucked away from the bedrock by the glacier's forward movement. Illus. R. J. Lethcoe.

Glacial Sculpting:

Glaciers moving downstream constantly pluck at sidewalls and valley bottoms. Evidence of GLACIAL PLUCKING is one of the most commonly observed phenomenon. Plucking of bedrock is probably due to strains on the rock. When the glacier presses against the uphill side of a bedrock protuberance, compression stress occurs. The greater stress melts some of the glacial ice. Meltwater then refreezes in pre-existing cracks in the bedrock's downhill side, expanding cracks and weakening the rock. At this point, the rock is partially incorporated into the glacier and partially still attached to the bedrock. Simultaneously, as the glacier flows over the downhill side of the knoll, it ceases to be fully in contact with the bedrock. Shear stress occurs on the downhill side of the rock. The combined effect of both the widening of pre-existing weak areas and the shear stress on the bedrock's downhill side can cause large boulders to be plucked away from the bedrock. Plucking also occurs when glaciers incorporate loose material into the ice. Large boulders are embodied in the glacier as the ice undergoes plastic deformation, surrounds and carries them along downstream (Fig-25).

Fig-26. Shear lines in the glacier indicate that the glacier has changed its direction of flow. Photo by author.

Fig-27. Striation marks are left by rocks embedded in the sole of the glacier scraping on bedrock. The criss-cross pattern of lines indicates periodic changes in the glacier's direction of flow. Columbia Glacier, eastern side. Photo by author.

the bedrock. Evidence of GLACIAL SCOURING is most easily seen on the smooth granitic rocks of Esther and Perry Islands or in the cirques and ridges occurring in the granite batholiths of Passage Canal, Port Nellie Juan, Eshamy, and Granite Bay. Excellent examples of glacial scouring can also be seen on the cliffs above the lake by Valdez Glacier and at many places in Thompson Pass.

Glacially worn bedrock, morainal material, and erratics often bear groove marks. Rocks in the glacier's sole produce grooves or STRIATIONS in other rocks (Fig-27). Sometimes the embedded rocks rotate in the ice presenting a new cutting edge and leaving striations which step sidewise. At other times, the glacier itself shifts direction leaving a trail of its course in the bedrock. The smoothed rocks on the cliffs near Valdez Glacier's terminal lake contain many good and conveniently located examples of striation marks.

CHATTER MARKS are crescent-shaped dents in the bedrock formed by the chip-

Fig-28. Chatter mark made in smoothed granite area by a boulder turning over at the sole of the glacier. Dark zone around chatter mark is lichen. Nellie Juan area. Photo by author.

An Observer's Guide to

Fig-29. Streams flow across the surface of glaciers, inside and at the bottom. Hikers on glaciers wear crampons on their boots, carry ice axes, ropes and other safety equipment. Photo by Nancy Simmerman.

CHATTER MARKS are crescent-shaped dents in the bedrock formed by the chipping action of larger rocks embedded in the sole of the glacier. They vary greatly in size from a few centimeters to several meters across. The pattern made by chatter marks varies with the kind of bedrock and the type of chipping rock (Fig-28).

Heavily silted streams carrying rock fragments flowing over, through and under the glacier erode variously shaped holes where they come in contact with the bedrock. In the sound, they seem most often to occur in granitic-type rock. Surficial geologists classify the holes according to the markings on the hole's side and their shape. PLUNGE-POOL POTHOLES are formed at the base of glacial waterfalls, have no spiralling curves engraved on their sides and are broad. Shallow depressions or GOUGE-HOLES in a stream bed are caused by spiral, circulation. EDDY-HOLES, usually the most common type, are very deep, sharp-edged holes whose walls are etched by spiral grooves left by rocks that have swirled around in the hole. Routes of former glacial streams can easily be traced by noting the contrast between rocks which have been finely polished by rapidly running water laden with glacial flour and adjacent rocks with a coarser polished surface abraded only by ice, meltwater, and ground rocks embedded in the glacier's sole.

Chapter 5. Glacial Landscapes

The Prince William Sound region owes much of its dramatic landscapes to former glaciers which carved out the fiords and valleys, horns and arêtes, truncated ridges and passageways, hanging valleys and sheltered coves. Today's glaciers continue this process.

Pleistocene glaciers transformed originally V-shaped stream valleys into glacially carved U-SHAPED VALLEYS. U-shaped valleys are formed by glacial erosion along their sides. When glaciers retreat, they leave a valley with overly steepened sidewalls and a rounded bottom (Fig-30). Subsequent deposition of outwash till and delta deposits may fill in the bottom creating a nearly level plain, as in Valdez Glacier Valley. Often the sidewalls remain unstable for centuries as rock and snow avalanches descend their sheer slopes. Some U-shaped valleys descend to sea level through a series of cliff and basin areas. Stairstepping glaciers (Fig-18) carved these valleys. Stairstepping and classic U-shaped valleys occur throughout Prince William Sound. Some of the most notable are in Passage Canal, Port Wells, Port Valdez, Port Fidalgo and Port Gravina.

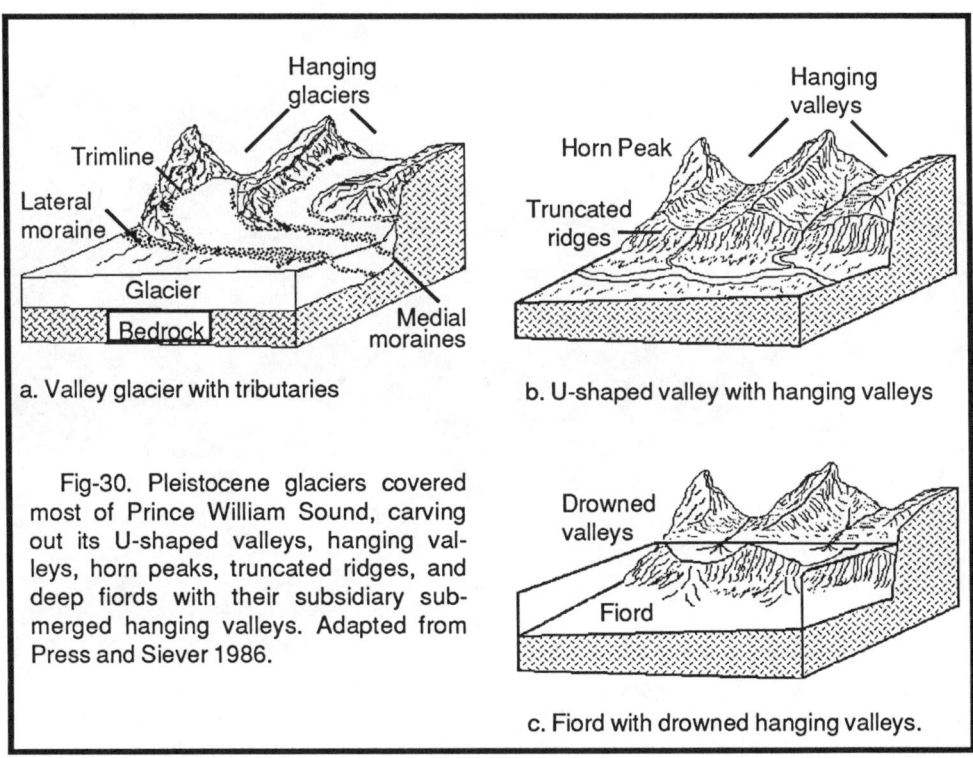

Fig-30. Pleistocene glaciers covered most of Prince William Sound, carving out its U-shaped valleys, hanging valleys, horn peaks, truncated ridges, and deep fiords with their subsidiary submerged hanging valleys. Adapted from Press and Siever 1986.

a. Valley glacier with tributaries

b. U-shaped valley with hanging valleys

c. Fiord with drowned hanging valleys.

In some valleys nearly horizontal ledges run along their walls. These may be strandlines left by former glacially-dammed lakes (Nellie Juan Lagoon), old lateral moraines (College Fiord), or weak areas in the bedrock (Passage Canal and Port Valdez).

TRIMLINES mark the former extent of the margins of a glacier. Sometimes a glacier advancing down a valley or fiord overrides forests in its path while leaving untouched forested slopes above its surface. After the glacier retreats, seeds from the remnant forest help revegetate the moraine-covered rock. However, a trimline remains for decades demarcating the unglaciated from the recently exposed area. Trimlines on both sides of Barry Arm are most interesting as they show the height of Barry Glacier during its most recent maximum around 1898 (Fig-31). The retreat of Columbia Glacier is leaving prominent trimlines between the glaciated and unglaciated areas.

HANGING VALLEYS form when larger and faster flowing valley glaciers erode their basins deeper than those of smaller tributary glaciers. The smaller, glacial valleys are left hanging high on the main valley wall. Scenic waterfalls often drop precipitously from the hanging valleys. Hanging, U-shaped valleys can be seen throughout the sound, but some of the best occur in Port Wells, College Fiord, Unakwik Inlet, and Port Valdez. A particularly beautiful hanging valley lies on the western side of Barry Arm about

Fig-31. The difference in vegetation still marks the trimline made by Barry Glacier during its most recent maximum, circa 1898. Coxe Glacier (pictured) separated from Barry Glacier, and this area became ice-free between 1909 and 1914. Photo by author.

The Glaciers of Prince William Sound, Alaska

Fig 32. Truncated ridges occur when a large glacier moving down a valley cuts across a ridge at right angles to it. As the glacier which occupied Port Nellie Juan during Wisconsinan time (late Pleistocene) moved past the southern end of Culross Island it simultaneously sculpted this truncated ridge and overrode Applegate Island. Photo by author.

one mile from Cascade Glacier where a waterfall plunges in stairsteps to Barry Arm over a 1000 ft. below. Its greatest single drop is 500 feet. This spectacular waterfall, which is visible from Port Wells, came into being with the 1898-1914 retreat of Barry Glacier. Ridges at the mouth of a hanging valley are usually truncated. TRUNCATED RIDGES occur when a large glacier moving down a valley cuts across a ridge.

Some hanging valleys carved by tributary glaciers that fed the sound's great fiord glaciers are now flooded bays. Comparison of the depths of side bays with those of the adjacent fiord reveal the underlying hanging valley type of topography. Thus, Pigot Bay hangs over Port Wells, Wellesley Cove and Yale Arm above College Fiord, Siwash and Jonah above Unakwik Inlet, Heather Bay above Columbia Bay, Shoup Bay above the depths of Port Valdez, and Fish Bay above Port Fidalgo.

FIORDS are U-shaped valleys flooded by salt water. They were created by the retreat of the massive tidewater glaciers which covered the sound and extended out into the Gulf of Alaska during the late Pleistocene. A common misconception is to think of the Pleistocene ice masses as solely on land. However, even if the sea level were 400 ft. lower, as it was at times during the ice ages, glaciers in this area were still scouring out their bedrock basins below sea level. Depths in the sound are considerably deeper than 400 feet; in some places, they reach to 2400 feet. The actual depth to bedrock is probably much deeper, since sedimentation partially fills the submarine valley floor.

The sound itself is a complex fiord system comprising a large outer fiord with many smaller, tributary fiords feeding it from three sides. Like all fiords, the sound and its

Fig-33. Horn peaks and serrated ridges called "arêtes" rise above the Chugach Icefield between Shoup and Columbia Glacier. Photo by John Weiland.

tributary fiords have one or more submerged bedrock sills. These are sometimes overtopped with morainal deposits. Sills separate the fiord's deeper basins from adjacent waters and sometimes segment the fiord itself. Water over these sills is much shallower than the adjacent fiord bottom. For example, the moraine bar bisecting Unakwik Inlet rises 700-750 feet above the fiord bottom. Sills cross all the passages between the sound and the Gulf of Alaska. Mariners with depth sounders can watch the precipitous rise of the fiord bottom as they approach and cross these sills and moraines.

Port Valdez is a typical glacial fiord; it has a barrier sill at the Narrows, a nearly flat bottom, and steep, glacially scoured sides. Hanging, U-shaped valleys above the fiord give some indication of the height of ice age glaciation — 3200 ft. Its originally rounded bottom has been leveled out by the deposition of tons of sediments. A rough estimate of the depth of these sediments can be gained from calculations based on the current rate of deposition. In 1973, Sharma and Burbank estimated that Port Valdez received 1.67 centimeters of suspended sediments annually plus an equal load in bedload sediments. Assuming for the sake of argument that this rate were constant over the centuries, then over a thousand feet of sediments would have been deposited in the past 10,000 years.

NUNATAKS are land forms, such as ridges or peaks, that are surrounded by a glacier. On a clear day, one can see hundreds of nunataks rising above the icefields of the high Chugach Mountains.

CIRQUES, ARETES, AND HORNS, sculptured by Pleistocene glaciers, abound in the Prince William Sound region. CIRQUES are formed by the bedrock scouring of

Fig-34. Roche moutonnée are formed by glacial plucking as the glacier flows over a bedrock knoll. Roche moutonnée range in size from a few to hundreds of feet high and are a common landscape feature in glacially sculpted areas. Illus. R. J. Lethcoe.

cirque glaciers. Two large cirques, one in sedimentary rock and one in granite, occur on the north side of Passage Canal. One of the largest and most spectacular cirques in Alaska lies at the head of Siwash Bay in Unakwik Inlet. Two contingent cirques make a saw-toothed ridge or arête. Some of the best examples of arêtes highlight the dramatic skyline above the Alyeska Tank Farm at Port Valdez. Both College Fiord and Port Gravina also have spectacular arêtes. HORNS are peaks formed by three intersecting cirque glaciers. Sugarloaf Mountain across the bay from Valdez is a classical example of a horn.

ROCHE MOUTONNEE refers to lumpy landscapes of polished and plucked bedrock surfaces. The mounds look very much like a fleet of overturned dinghies with their bows pointed uphill. Examination of the smoothed, uphill side of the exposed bedrock often reveals striation lines. The lee side, which looks like the blunt transom of a dinghy, may have piles of rock rubble left at its base. Vegetation gains an early foothold in these areas. Examples of roche moutonnée topography can be seen at the Nellie Juan Glacier area, in Icy Bay, along the western side of College Fiord, and at Shoup Bay.

Tarns and many lakes owe their origins to glacial sculpturing. TARNS occur where glaciers scour deeper into a bedrock ridge creating small hollows. Subsequently, these fill with rain and snow meltwater. Many tarns do not have outlet streams. Cirque basins often contain lakes, such as those on the western side of Culross Island. Long, narrow, lakes whose outlets flow through old terminal moraines or bedrock ridges may fill the floor of former valley glaciers. Esther, Surprise, Solf (Knight Island), and Solomon Gulch (Port Valdez) Lakes represent this type.

GLACIER-DAMMED LAKES form behind glaciers in side valleys where the main glacier blocks direct outflow. Both advancing and retreating glaciers can form glacially dammed lakes. Ice-free tributary valleys blocked off by glaciers have the largest glacier-dammed lakes. Small lakes nestled between the main glacier and a retreating glacier, such as Camicia Lake (now drained) behind Valdez Glacier, or in niches in valley walls are the most common. Most glacially dammed or "dumping" lakes drain subglacially or en-glacially; some cause sudden, catastrophic floods called "jokulhlaups" when they breach their ice dams. The Richardson Highway through Keystone Canyon has been repeatedly threatened and destroyed by glacial outbursts from lakes formed by two glaciers in the Sheep Creek Basin. "The section of the Valdez-Fairbanks Trail through Keystone Canyon [Richardson Highway] . . . was . . . one of the most expensive stretches to maintain in

Alaska. High water, often caused by bursting of glacier reservoirs, annually required expensive maintenance in Keystone Canyon. During the summer of 1913, the bridge on Sheep Creek was carried away by a flood caused by the bursting of a glacier reservoir at the head of the creek. . . In 1916, a glacier reservoir that burst at the source of a small stream at the head of Keystone Canyon required the reconstruction of that section of the road. In 1919, Bear Creek at Mile 18 filled its channel with 20 feet of boulder, gravel and debris, destroying the bridge." (Post and Mayo 1971, quoting Hoffman 1970, p. 36). Outburst floods sometimes combined with heavy precipitation have destroyed bridges in this area again in 1945, 1959, and 1981.

STRANDLINES, created by lake wave action and by deposition from the heavily silted lake waters on a hillside, indicate the presence of a former glacially dammed lake and its height between dumpings. Strandlines can be seen in the indentations on the east and west sides of Nellie Juan Glacier. From a distance they appear as successive terraces. The different strandlines provide a glimpse into the glacial history of the area, since each one indicates a filling of the lake between dumpings and hence possibly different levels of the blockading glacier.

Fig-36. Columbia Glacier has five ice-dammed lakes: Terentiev, Kadin, Chaos, Borealis, and Number One. Terentiev appears in the foreground and Kadin in the narrow valley behind. As Columbia retreats, Chaos, Borealis, and Number One will probably disappear. Kadin and Terentiev may drain over bedrock sills. Moraine patterns show how Columbia sometimes flows into Terentiev and Kadin, then backs off. Terentiev can best be viewed by hiking up the ridge on the west side of Columbia Glacier. A. Post, USGS. Sept. 1981.

Chapter 6. Glacial Drift

Glacial erosion plays a vital role in transforming bedrock and boulders into smaller particles capable of releasing their rich mineral content into the soils. Time and again, Prince William Sound's glaciers have advanced covering existing forests, then retreated leaving behind pulverized rock debris from which new, revitalized soils slowly form. The process takes time, since not all minerals necessary for plant life are found in glacial rock debris. Nitrates, for example, are not available. And, rock debris is not soil. Soil is composed of both mineral and organic matter.

An afternoon's walk near a glacier provides many opportunities to observe the effects of various types of glacial erosion. As glaciers flow, entraped rocks erode bedrock surfaces by plucking and polishing. Erosion caused by glacial streams also contributes to the variety of past and present features readily observed in the area.

All glacially deposited material found on land or sea is called DRIFT. Till and outwash are two types of drift. TILL is mixed rock debris which has been deposited by a

Fig-36. Icebergs from tidewater glaciers can transport glacial erratics and drift long distances from the glacier's terminus. Photo by author.

Formation and development of an Outwash Plain

Fig-37a. An active outwash plain with braided streams. A large block of ice has been stranded by the retreating glacier. Adapted from Press and Siever 1986.

Fig-37b. An old outwash plain whose rivers have become stablized. The depression left by the stranded block is now a kettle and shrubs grow along the riverbanks.

glacier, such as in an ablation moraine; it has not been sorted by extensive stream action. Till deposited by icebergs may be a long distance from the glacier's terminus. OUTWASH is drift which has been carried and sorted by stream action. A hole dug into till reveals a poorly sorted accumulation of silt, clay, pebbles, and sand, while a hole dug into outwash shows an upper layer of clays, lower bands of cross-bedded sand and gravel, and finally a lower level still of pebbles. This banding of larger rock debris at the bottom and finer ones at the top is typical of water transported and deposited sediments.

OUTWASH PLAINS are gently sloping lands in front of terminal moraines. Numerous streams and abandoned stream beds dissect the outwash plain. While stream courses

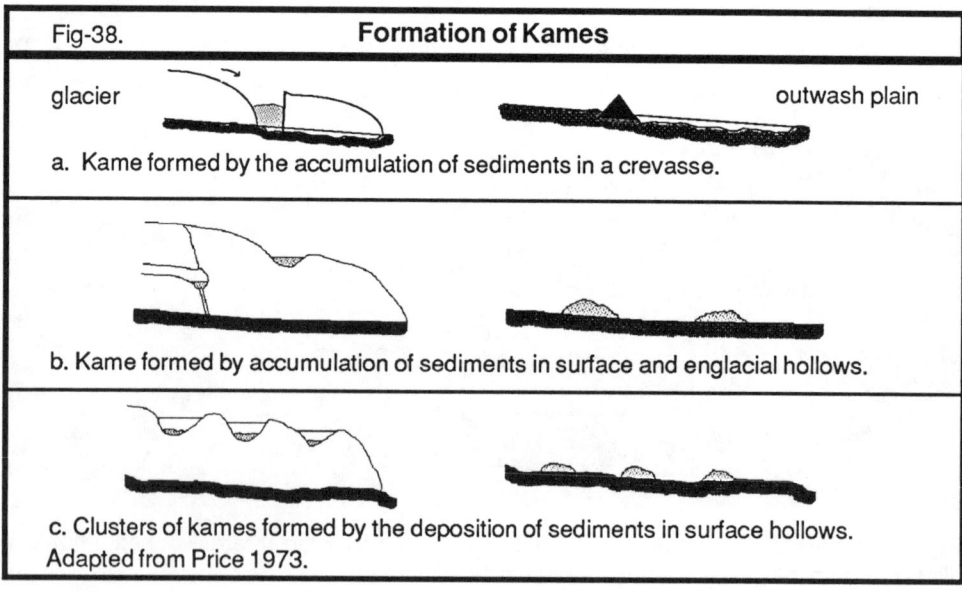

Fig-38. **Formation of Kames**

a. Kame formed by the accumulation of sediments in a crevasse.

b. Kame formed by accumulation of sediments in surface and englacial hollows.

c. Clusters of kames formed by the deposition of sediments in surface hollows. Adapted from Price 1973.

fluctuate and the plain is being actively constructed, little or no vegetation grows. What does grow is subject to glacial floods and washouts. Once the stream establishes its course, riparian woodlands dominated by alder, willow, and black cottonwood develop.

A typical outwash plain may contain glacially deposited till such as erratics, kames, eskers and moraines. When the place of origin is different from the place in which a rock is deposited, the rock is termed a GLACIAL ERRATIC. Glacial erratics ranging in size from pebbles to boulders 20 or more feet high are abundant in the region. The terminus of Nellie Juan Glacier cuts across a granitic stock, yet many of its morainal rocks are sedimentary erratics carried from sedimentary peaks high in the Sargent Icefield.

Streams running across the surface of glaciers carry and grind small rocks in their course. Eventually, the streams plunge from the glacier's surface through a crevasse or MOULIN to its base. If the glacier's position remains constant for a long period, semi-polished rocks are carried down the stream and over the waterfall forming an inverted cone-shaped pile called a KAME. Kames also form along the margins of a glacier where sediments accumulate in ponds and areas where blocks of ice have become separated from the glacier (Fig-38). Kames are typically found in kettle-type topography. Stones found in a kame are generally more rounded than morainal debris and less rounded than rocks from englacial streams, rivers and deltas.

KETTLES are formed by large blocks of moraine-covered ice stranded on an outwash plain. While the ice slowly melts, rock debris from the glacier continues to build up around the ice block. When the block finally disappears, sometimes after many years, a depression remains which may either fill with water or remain a dry hole. Kettles have formed near the margins of both Sheridan and Shoup glaciers. Old kettles occur on Barry Arm's former outwash plain across from Pt. Doran, in front of Taylor Glacier and to the east of Unakwik's morainal bar (where they are surrounded by spruce forests).

ESKERS are meandering ridges of stream-worn gravel deposited by a meltwater stream flowing in or under nearly stagnant ice. Often the winding, serpentine-like ridges of rock cross depressions and hills, indicating the route taken by the englacial stream through the ice to its outlet (Fig-37b).

Outwash plains can be seen by Baker and Toboggan glaciers in Harriman Fiord and in front of Bettles, Tebenkof, Ultramarine, Nellie Juan, Taylor and Princeton glaciers. Near Valdez, the Richardson Highway parallels the outwash plain of the Lowe River, where there are excellent examples of a heavily braided drainage system and the instability of the stream channels. Repeated flooding by the Lowe and Valdez Rivers has played havoc with the highway and Old Town Valdez since their founding.

GLACIAL STREAMS carry heavy loads of sediments. Sediments, including boulders, moved by sliding, rolling or bouncing along the bottom are called BEDLOAD. Smaller sediments, such as glacial flour, that remain in suspension in the stream are called SUSPENSION LOAD. The levels of glacial streams fluctuate annually as melting is slowed or stopped in the winter and increases in the summer. Studies of glacial streams entering Port Valdez show that stream-flows peak in June, as a result of snowmelt, and again in August and September, as a result of rain and increased glacial melt (Sharma and Burbank 1973). Because rocks in an outwash plain are unconsolidated, stream banks are easily washed away by the glacial stream itself and sediments carried in the stream. Streambank erosion increases during and immediately after periods of heavy rainfall, when rivers become swollen with runoff and increased glacial meltwater. Often, during an autumn gale, one can stand near a stream and listen to the knocking and pounding of boulders being bounced along the bottom.

When an outwash stream overflows its banks, it leaves a fan-shaped deposition of gravel, sand and silt called a SANDUR (plural: sandar). Catastrophic floods from glacially dammed lakes have very high discharge rates which can cause sudden and unexpected stream bank erosion, changes in channels, and deposition of large volumes of boulders, cobbles, gravels and sand in a sandur. During one such flood on Sheep Creek (Richardson Highway near Valdez), 25 feet of rock debris, including many boulders, was deposited (Post and Mayo 1971).

DELTAS form when streams deposit their sediments in a lake or marine environment. Because of the area's steep slopes, high seasonal runoff, and high rate of erosion both bedload and suspension load sediments are carried to the mouths of the streams where they build deltas. Here, the river current, slowing on its encounter with the sea, first drops coarser sediments near shore, then medium and finer sediments farther offshore. Sediments deposited closest to shore on a horizontal level are called TOPSET BEDS. Sediments deposited on the incline form FORESET BEDS, while those that settle to the bottom of a fiord are called BOTTOMSET BEDS. Finer grained sediments that settle to the bottom may have first been carried several miles from the mouth of the stream (Fig-40).

Over the centuries, deltas slowly build up until saltwater-tolerant plants are able to invade establishing salt marshes. Plant roots trap the finer sediments causing them to settle and raise the land even further. Eventually, plants less tolerant to salinity invade the marsh transforming it into a beach meadow community.

Tectonic events, such as the 1964 Earthquake, have contributed to the abrupt uplifting of some deltas and submergence and massive sliding of others. The Copper River Delta, one of the largest deltas in the world, uplifted during the earthquake. Uplift caused major changes in the plant communities, as land plants such as alder spread over the marshland. As a result, female brown bears with cubs finding good cover in the alder shrublands moved into the marshlands and began feeding on the eggs and young of nesting waterfowl. By contrast, Old Town Valdez, located on the Valdez Glacier delta, had part of its waterfront suddenly slide into the sea. After the Earthquake, the town had to be relocated 4 miles west to the Mineral Creek delta. Here, deltic sediments are held in place by a series of glacially scoured, bedrock knolls.

Chapter 7. History of Prince William Sound's Glaciation

Pre-Pleistocene Geology and Glaciation:

The Earth's climate has changed dramatically and periodically during her long history as witnessed by Pre-Cambrian (670 mya = million years ago) and Permo-Carboniferous (290 mya) glaciation in other parts of the world. Geologists have identified other ice ages dating at 2300, 950, 770, and 440 mya. These would have had little effect on the Prince William Sound region: the sediments composing its rocks did not start forming until about 180 mya.

From the Triassic Period (245-208 mya) to early and mid-Tertiary Periods (67-1.8 mya) world temperatures remained remarkably constant and much warmer than todays. In the early Tertiary Period (40-63 mya), a mild undifferentiated climate prevailed worldwide. During this same period, the Chugach and Prince William Terrane (part of a former plate) moved north from its area of origin, possibly off Baja California, to dock with Alaska about 60 mya. Following the terrane's occupation of its present position, climatic conditions changed. Perhaps by the middle of the Oligocene Epoch (25-40 mya), the Earth's climate cooled, and regional climates developed.

Why did the climate change? One possible answer is that when the North American and Eurasian continents finally reached their present positions many tens of millions of years ago, new oceanic and atmospheric weather and heat transfer patterns developed. These were accentuated by Antarctica splitting off from South America and moving south where it was cut off from the warm equatorial waters. As the temperatures cooled and a more varied climate developed, glaciers gradually formed in Antarctica (Covey 1983).

The Earth's orbit may also play a role. Some astrophysicists and glaciologists have noted a connection between variations in the Earth's attitude in space and the shape of its orbit (circular or elliptical) with the onset of more recent ice ages (Calder 1974; Covey 1983). At optimum times, the sun appears higher in the northern summer sky due to the position of the Earth's orbit. Its rays are concentrated in a smaller area; the climate warms, glaciers retreat. When Earth's orbit shifts, the sun hangs low in the northern sky dispersing its weak rays over a larger area; the climate cools, and glaciers advance.

Miocene and Pliocene Glaciation:

The relationship between worldwide ice ages and specific ice advances in the Prince William Sound region during the Miocene (25-6 mya) and Pliocene (6-2 mya) periods remains obscure. The Chugach and Kenai Mountains have probably been continuously glaciated since about 6.3 mya (Molnia 1986). Evidence on land, such as moraines, for the earliest glacial advances has been obliterated by subsequent advances.

When glacial geologists look at core samples of oceanic sediments, they often find layers with large amounts of ice-rafted sediments. The presence of ice-rafted sediments indicates that at the time the sediments were formed the glaciers were tidal. Recently, glaciologists have turned to studying sedimentary bands in The Yakataga Formation, a rock unit which stretches from Middleton Island to La Perouse Glacier, just south of Lituya Bay. At the formation's eastern end, they have found ice-rafted, glacial deposits in Miocene age sediments. However, because there are no Miocene age rocks in the Yakataga Formation at its western end, there is no record of Miocene glaciation here (Molnia 1986).

It is possible that during this period the Chugach Mountains were not as high as they are today, so glaciation may not have been as extensive. Furthermore, at this time Prince William Sound was probably a coastal plain (Tarr and Martin 1914; Molnia 1986). Glaciers flowed out onto the plateau and were not yet tidal.

Sediments dating from between 2 and 3 mya (Pliocene) from the eastern portion of the Yakataga Formation contain large quantities of ice-rafted sediments once more indicating that glaciers either extended out onto the continental shelf or were near shore (Molnia 1986).

Pleistocene Glaciation:

During the early Pleistocene Epoch (1.8 mya) the Earth's temperature again cooled causing severe climatic changes. Permafrost spread, and the world's ice-cover formed. Periodically, the ice-cover expanded and contracted. During times of expansion, large ice sheets covered southcentral Alaska, most of Canada, the northern third of the contiguous United States, much of Europe, all of Scandinavia, and large parts of Siberia. Northern Alaska, the southern half of North America, Central America, and the northern half of South America were ice-free. Small, isolated pockets of ice-free land may also have existed on the western side of Vancouver Island, the Queen Charolette Islands, on some of southeastern Alaska's islands and possibly Cape Yakataga. In the Prince William Sound region, glacial geologists noted signs of ice-free areas on the upper elevations of Hinchinbrook (above 400 ft.), Montague (Tarr and Martin 1914), and Knight (Grant and Higgins 1913) Islands and possibly on portions of the Copper River Delta. These isolated island nunataks probably supported the plant life that furnished the seed source for the Holocene revegetation of the sound (Heusser 1955, 1983).

The most extensive ice ages saw 30% of the world's land-base covered by ice. Water stored in the ice caused the sea level to drop a maximum of 400 hundred feet, periodically opening up the Bering Land Bridge (Beringia) between Alaska and Asia. Because the continental glaciers blocked the interchange of plants and animals with the rest of North America, while the land bridge facilitated interchanges with Eurasia, Alaska was at these times virtually part of Siberia. Plants, animals, and eventually, man moved across Beringia. The climate between ice ages was warmer than today's; ice retreated to cover less of the earth's surface than it does now; and the sea level rose submerging coastal areas. Beringia flooded, opening up a waterway between the Atlantic and Pacific oceans and closing the land connection between Alaska and Eurasia.

Little research has been done on Pleistocene glaciation in Prince William Sound. Again, we must look to studies of the eastern Gulf of Alaska for suggestions as to what might have been occurring here. Sometime between 1.8 and 0.7 million years ago, copious amounts of glacial sediments were deposited in the Yakataga Formation on Middleton Island, suggesting the likelihood of coastal marine glaciers. The sound's fiords, the surrounding glacially sculptured landscape, and the bathymetry of the adjacent Gulf of Alaska all clearly indicate that glaciers covered and scoured this area for thousands of years. Glaciologists assign this extensive glaciation to the Wisconsinan time, the last of several Pleistocene Ice Ages.

Tarr and Martin (1914) were the first glaciologists to describe the probable course of glacial events leading to the creation of Prince William Sound. During the Wisconsin Glaciation (late Pleistocene), icefields formed in the high Chugach and Kenai mountains, then flowed down pre-existing valleys to a wide plain which filled the interior part of pre-glacial Prince William Sound. Glacier tributaries flowing from the mountains joined into two main streams as they crossed the plain: one stream flowed from Port Valdez and Columbia Glacier out Hinchinbrook Entrance. The other major stream united glaciers from northern and western Prince William Sound and flowed south, separating into two streams as it pressed around Knight Island whose higher peaks stood as nunataks above the glacier, then reunited to flow out Montague Strait. Smaller glacial streams flowed out through Latouche, Elrington, Prince of Wales and Bainbridge Passages.

It is not known how far offshore the glaciers flowed, as no terminal moraines have been discovered south of Prince William Sound, although some offshore moraines have been located on the continental shelf from the Alsek River to Icy Bay. However, striation marks on rocks at Middleton Island indicate that at least one ice-tongue extended that far south sometime during the middle to late Pleistocene (Molnia). A large trough extends southward from Hinchinbrook Entrance on the eastern side of the sound. This trough has the typical U-shaped valley profile and tributary side troughs (hanging valleys) so indicative of glacially carved valley and fiord systems that it was undoubtedly carved out by an ice-tongue advancing out Hinchinbrook Entrance (Carlson 1982). Just how often and how far onto the continental shelf glaciers originating in Prince William Sound spread during the late Pleistocene is unknown.

Fig-39.	Height of Wisconsin Ice Age Glaciers		
Location	Feet above sea level	Source	
Hinchinbrook Ent.	400 ft.	Tarr and Martin	
Orca Inlet	2,300 ft.	Tarr and Martin	
Montague St.	2,900 ft.	Moffit	
Latouche Island	2,000 ft.	Tarr and Martin	
Knight Island	2,400 ft.	Grant and Higgins	
Ellamar	3,000 ft.	Capps	
Columbia Bay	4,000 ft.	Tarr and Martin	
Port Valdez	3,200 ft.	Grant and Higgins	
Harriman Fiord	4,000 ft.	Tarr and Martin	
College Fiord	4-5,000 ft.	Tarr and Martin	

It is difficult to determine the total thickness of the ice mass, because glaciers, like rivers, varied in depth from one place to another. In addition, the land, which was depressed by the weight of the glacier, has rebounded differentially. However, it should be remembered that even after the Wisconsin glaciation carved the land down to below sea level, tidewater glaciers remained in contact with the bedrock at their sole. Thus, if we add the height of the glacier above water to the depths scoured out below sea level, we can reach an approximate figure for the ice stream's thickness. For example, near Knight Island the glacier's height was 2400 feet above sea level, and its depth below sea level as indicated by NOAA chart 16700 was 2400. At the very least the ice was around 4,800 feet thick. It was probably much thicker, since the depth of the water is not the same as the depth to bedrock, which may be covered by hundreds of feet of sediments.

The weight of this mass of ice depressed the earth's crust approximately 1/6 to 1/10th the amount of the ice-cover. For example, if the ice was 4800 feet thick at Knight Island, than the Earth's crust was isostatically depressed between 480-864 feet. When the ice-cover melted, isostatic rebound occurred. The exact amount of the crustal uplift cannot be determined because Prince William Sound is a tectonically active area; earthquakes have alternately raised and lowered the land considerably over millions of years.

While glaciers were carving out the sound, other glaciers flowed north from the Chugach Mountains filling the Copper River Basin and fronting on a glacially-dammed lake. A large ice-dam on the Copper River, as it flowed through the eastern Chugach Mountains, created the lake (Williams and Johnson 1979).

The Holocene Period: the last 10,000 years

By 10,000 to 12,000 years ago the coast was probably ice-free. Radiocarbon dates on duff layers near bedrock show evidence of plant life at Golden (mouth of College Fiord) around 10,150 years ago (Heusser 1983), Perry Island about 9,440 years ago (Heusser 1983), Port Valdez roughly 9,520 years ago (Williams and Coulter 1979), and Alaganik (near Cordova) about 10,390 years ago. The Kenai Peninsula to the west was deglaciated and invaded by plants by 14,000 years ago (Heusser 1955). To the southeast, Glacier Bay became ice-free about 13,000 years ago, and Controller Bay (behind Kayak Island) was ice-free 14,000 years ago (Sirius and Tutelary 1969). Glaciers flowing northward into the Copper River Basin probably began retreating between 11,000 to 12,000 years ago; by 8,000 years ago most occupied positions near their present termini. The glacially-

Fig-40. **Development of Mineral Creek Delta, Valdez Townsite**

a. 15,000 to 12,000 yrs. ago a glacier fills Port Valdez and possibly extends out through Hinchinbrook Entrance on to the continental shelf. Sea level is about 400 ft. lower than present.

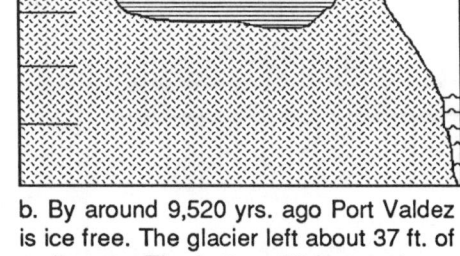

b. By around 9,520 yrs. ago Port Valdez is ice free. The glacier left about 37 ft. of sediments. The bottom 26 ft. are clayey silt overtopped by 11 ft. of dense sand and gravel. A freshwater marsh has formed. Sea level is about 240 ft. lower than present.

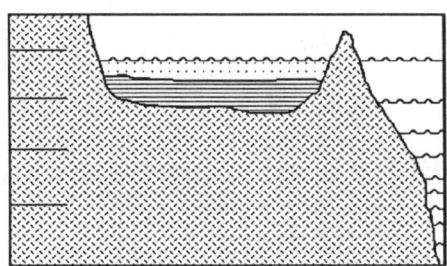

c. By 6,000 yrs ago the sea level is about 65 ft. below present levels. Water carried glacial bedload and suspended sediments are deposited on the delta. Isostatic rebound (rising of the earth's crust) and eustatic (world-wide changes in sea level) continue to occur.

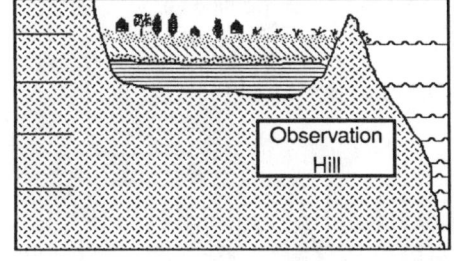

d. 1,000 to present: Water-carried glacial sediments have been deposited in bottomset beds (6 ft. deep), foreset beds (28 ft. deep) and topset beds (25 ft. deep). Tectonic events continue to raise and lower the land. Parts of the delta are above sea level.

Based on Shannon & Wilson, Inc. 1964; Williams & Coulter 1979; Molnia 1986.

dammed lake drained through the Copper River Canyon sometime between 9,000 and 10,000 years ago (Williams and Johnson 1979).

Melting of the Wisconsin Age glaciers not only revealed a new, glacially sculpted landscape, but also caused a significant rise in the sea level between 8,000 to 10,000 years ago. Borehole samples of coastal glacial outwash plains, such as the Mineral Creek delta (site of new town Valdez), show that a freshwater bog developed first after the glacier's retreat. This was subsequently submerged by the rising sea level. Stream borne glacial sediments gradually filled in the delta until the land rose above sea level [Fig-40].

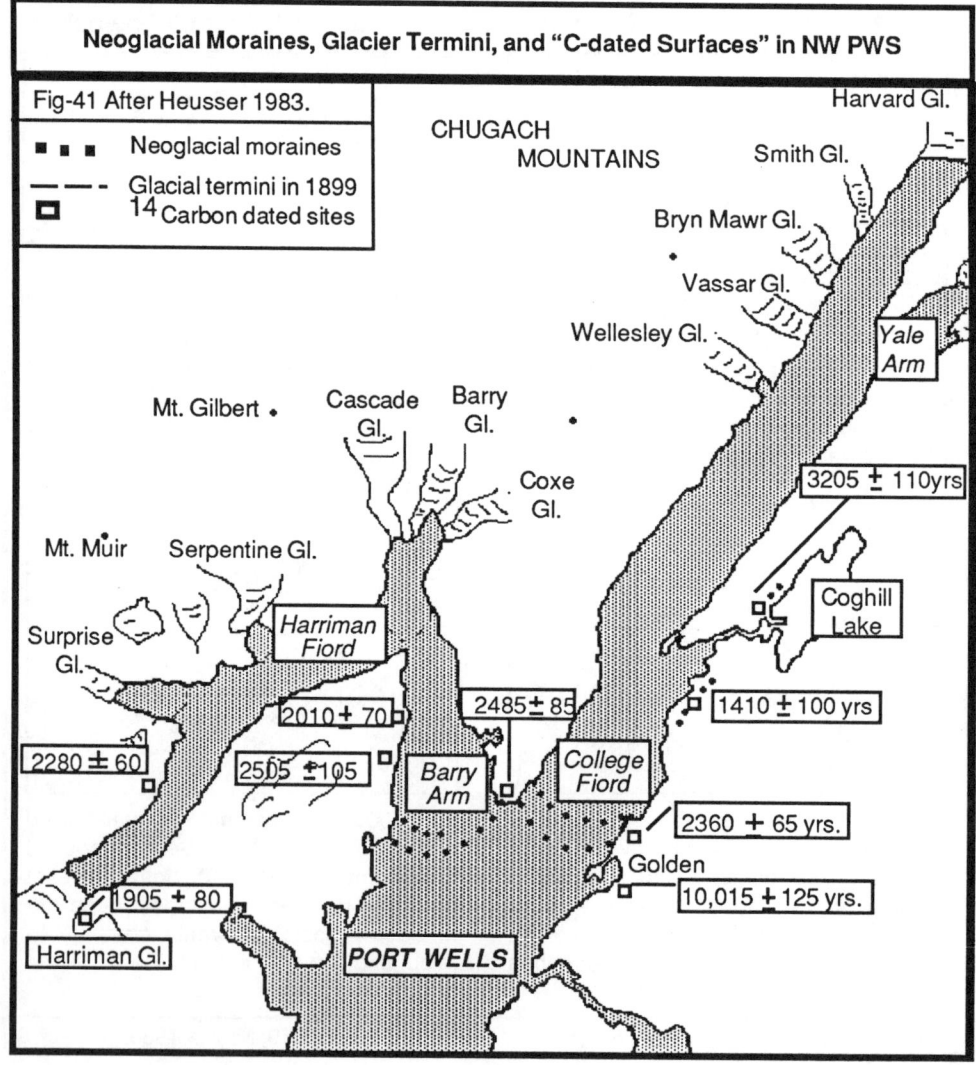

Very little study of the sound's outwash plains and deltas has been done outside of the Valdez area. Test holes made during Alaska Pacific Refinery's study of the area for a proposed refinery and petrochemical development in the Valdez Glacier Valley reveal that in places the bedrock has been covered by 700 ft. or more of drift. Some evidence was found suggesting morainal deposits from earlier glacial advances and silts indicative of former deltas (ALPETCO 1979).

The most important study for establishing neoglacial activity has been Heusser's use of radiocarbon dating combined with his interpretation of pollen and sediment records from borehole sites in northern and northwestern Prince William Sound. Heusser (1983) found that a coastal sedge tundra community interspersed with thickets of willow and alder was typical of periods with a cool, moist climate. Coastal sedge tundra communities occurred as early as 10,015 years ago at Golden (near Coghill in College Fiord). Subsequent dominance of alders (8300 to 3300 years ago) suggests a warmer period. Between 3200 and 2500 years ago, alder populations declined and were replaced by sedge tundra communities interspersed with small pockets of mountain hemlock, Sitka spruce and western hemlock on well-drained soils. Vegetative evidence thus indicates a cooler, neoglacial period at this time. The advance of conifers after 2000 years ago indicate the end of the neoglacial period and the replacement of alders by a new, warm period, dominant species — the conifers.

A more recent decline in the conifers and advance of sedge tundra communities accompanied the last neoglacial period (Heusser 1983). Within the past few decades, many of the sound's lowlying peatland bogs and ponds have again dried out in response to the climate's warming following the end of this most recent neoglacial period. Warmer temperatures mean less precipitation falls as snow in the lower elevations. Snow-cover, which remains longer over the bogs than in the forests, now generally melts between April and early June (depending upon exposure), whereas a decade or two ago it lasted into July and even mid-August.

Thus, although the Holocene has been generally warmer than the Pleistocene period, this warming has not been steady. The earth's climate and glaciers have continued their fluctuations. When the climate cooled and precipitated more moisture, glaciers waxed. In warmer, drier periods, they waned. World-wide, there is also evidence for several major neoglacial periods (Little Ice Ages) during the last 6,000 years. Today, many of the sound's glaciers, such as Valdez, Toboggan, Cataract, Ultramarine, Tebenkof and Claremont, are retreating — probably as a result of the recent warming trend. If the trend continues, some of the smaller apron and cirque glaciers may disappear. However, because tidewater glaciers respond both to climate and the position of their terminus, advances and retreats do not necessarily correspond to neoglacial periods as demonstrated by the present advance of nine tidewater glaciers.

Chapter 8. Explorations and Scientific Investigations

Most early European explorers of Prince William Sound were primarily navigators who knew and cared very little about natural history. According to Davidson, "In his (Capt. Cook's) second and third voyages he had no scientific men with him. Vancouver was with him in both these voyages, and would appear to have accepted Cook's view that naturalists were "disturbers of the peace"; a dictum that reaches farther back, and has not disappeared to the present day." (Davidson 1904, p. 5). Consequently, maps by Cook, Vancouver, and Teben'kof show increasing accuracy concerning the shoreline but do not indicate other features such as glaciers. Occasionally, their journals mention ice in the fiords; however, it is difficult to know if they are describing the position of glaciers, sea ice, or mirages.

The Russian engineer Doroshin, who spent from 1848 to 1858 surveying Russian America for minerals, notably gold, appears to be the first person to find the sound's glaciers intriguing in their own right and to speculate on the causes leading to their formation. Doroshin was particularly bemused to find that while mountains near St. Petersburg and Stockholm had no glaciers, in Alaska places with the same latitude had numerous glaciers. He discusses in detail then current theories regarding the formation of glaciers. In light of what he has seen in the sound and elsewhere in Alaska, Doroshin rejects the then current view that glaciers form only where the height of the mountains is above the snowline or at certain latitudes. Instead, he argues, correctly, for the theory that glaciers form where the annual accumulation of snow exceeds melting.

Almost half a century passed before the sound's glaciers were again visited by men of intellectual curiosity. While inventorying fisheries resources for the Alaska Fisheries Commission in 1887, Samuel Applegate marked the location of several glaciers on his map, thus becoming the first person to map the glaciers' locations. Unfortunately, he did not leave a written record, and his map was not widely circulated (Davidson 1904). Abercrombie, for example, appears not to have known about it, since his own 1899 map omits bays, such as Kings Bay, discovered by Applegate.

The best and most widely publicized of the new explorers were American Army personnel, who were well-educated and trained to notice and appreciate detail. Of this new wave of explorers, Lt. Abercrombie was one of the finest. His journals are full of observations on plants, animals, geology, and, of course, glaciers, as his notes on Childs Glacier indicate:

> . . . I beheld one of the grandest spectacles ever witnessed by living men. First, bergs came majestically sailing down stream, passing and repassing each other as the force of the water backed up by their united

presence became sufficient to force them through the sandy bottom of the river, which they in turn ground to a find composition, similar in color and consistency to the deposit found under a grindstone after the sharpening of tools. These monsters differ in color. Some are white, others black, and yet others (the latter predominating) are of an aquamarine color. As they backed and filled they somewhat resembled the maneuvers of a fleet of men-of-war. By observation I estimated the discharge of Childs Glacier to be as follows: The breadth of this glacier as cut by the river, which runs at right angles to the flow of this field of ice, is a trifle over 3 miles; its mean altitude is 400 feet; discharge per day, from June 1 to August 31 (ninety-two days) 3 inches, giving over a face of 336,000 square feet, and 23 feet drop, 8,160,768,000 pounds per annum. (Abercrombie 1900, p. 388.)

The 1898 Army reconnaissance included two U.S. Geological Survey scientists, W.C. Mendenhall and F.C. Schrader. In the course of their other work, both men sometimes photographed, described or mapped the termini of glaciers, but they were naturally more interested in access to the interior and the glaciers, such as Portage and Valdez, which provided it. Schrader summed up his reconnaissance of Valdez Glacier thus:

From a topographic standpoint the trip was a very material success, as it for the first time determined the size and trend of Valdes Glacier, the character of the country in which it lies, the altitude of the summit and surrounding mountains, and the nature of the country at the head of the Klutena River on the inland side of the divide. (Schrader, p. 353).

Over sixty years was to pass before the upper reaches of a second glacier in the region was mapped in detail from the ground. The credit goes to Lawrence Nielsen, a glaciologist and mountaineer, who explored and mapped the upper reaches of Columbia Glacier in 1962 (Nielsen 1963).

Abercrombie's map (1899) indicates and names Childs, Miles, Sheridan, Sherman, Valdez, Shoup, Columbia (Live Glacier), Harvard and Yale (Twin Glaciers). He maps unnamed glaciers at the head of Unakwik Inlet (Meares Gl.), Barry Arm (Barry and Cascade Gl.), along the west side of Port Wells, Passage Canal (Portage Gl.), and Icy Bay (Tiger Gl.). His map does not show Harriman Fiord or Kings Bay.

In 1899, the railroad baron, Edward Henry Harriman, undertook a private cruise for his health; for company, he invited the nation's most eminent scientists to explore Alaska's coastline. Both John Muir and G. K. Gilbert took this as an opportunity not only to refine their theories on glaciers (Meier and Post 1980), but also to establish baseline data on the glaciers which could be used by future glaciologists in studying glacial movement. By this time, the study of glaciers had shifted from theory to long-term study and documentation. Gilbert was the first person to recognize the importance of establishing photographic stations marked by cairns and indicated on maps so that future

An Observer's Guide to

Fig-42. Edward H. Harriman chartered the steamer *George W. Elder* to take his family, friends and guests on a hunting and scientific exploration trip along Alaska's coast. Here, they explore College Fiord (Wellesley Glacier in background) just prior to their discovery of Harriman Fiord. The overall scientific achievements of the Harriman Alaska Expedition were not matched for depth and breadth by any other group until 1964, when following the Earthquake scientists from many disciplines came to study its effects. Merrian Photo. Permission to publish granted by the Arctic and Polar Research Library, U. of A.

scientists could return to the same location and photograph the changes in the glacier. They photographed and made the first maps of the termini of Columbia, Amherst, Crescent, Holyoke, Vassar, Bryn Mawr, Wellesley, Smith, Harvard, Yale, Barry, Serpentine, Toboggan, Surprise, Cataract, Detached, Wedge, Roaring, and Harriman Glaciers. In addition, they discovered and mapped Harriman Fiord itself. Scientific papers from the expedition were published in thirteen volumes frequently referred to as the *Harriman Alaska Expedition*. The importance of this privately funded research cannot be over-estimated. It was not until the 1964 Earthquake that another similarly diversified group of scientists returned to the sound to study its various geological and biological aspects in detail.

In 1905 and 1909, Grant and Higgins of the U.S. Coast and Geodetic Survey, made a more complete endeavor to survey the sound's glaciers reoccupying photographic positions established by the Harriman Expedition and visiting glaciers, such as those in Unakwik, Port Nellie Juan, and Icy Bay, that had been skipped by the Harriman Expedition. They, too, established photographic stations. Grant and Higgins' work was first published in the *Bulletin of the American Geographical Society* in 1910 and 1911. A full report was published by the U.S. government in 1913. Their work and that of Tarr and Martin in 1910 (*Alaska Glacial Studies 1914*), sponsored by the National Geographic Society, mapped the termini and general orientation of the glaciers and continued establishing baseline data for detailed, long-term study of their behavior.

Unfortunately, after this early rush of exploration, only a handful of glaciologists visited the sound during the next fifty years. William O. Field, who made eight glacier

Fig-43. Bradford Washburn made the first professional aerial photographs of Prince William Sound's glaciers thus opening up a whole new range of glacier studies. Shoup Glacier is in the foreground with Columbia Glacier in the distance. Shoup's terminus is well established on the moraine where it remained until the early 1960s. Note the plume of glacial silt trapped behind the outer moraine. Photo taken June 14, 1937.

Fig-44. William O. Field in 1976 in Cordova preparing for one of his trips to survey the area's glaciers. Field's research on Prince William Sound's glaciers spans fifty-six years (1931-1987). (Photo courtesy of William O. Field.)

surveys in the sound between 1931 and 1976, is the outstanding exception. His work, *Mountain Glaciers of the Northern Hemisphere,* is a leading authority on Prince William Sound's glaciers. While collecting data on glaciers, Field made a number of important observations, including the fact that a few tidewater glaciers in College Fiord appeared to have periodic advances and retreats (Field 1975). Field's photographs and detailed collection of data concerning a large number of glaciers helped glaciologists acquire the information necessary to begin understanding glacial dynamics. For example, Post and Meier built on Field's observations of the behavior of tidewater glaciers when developing their theory of calving glaciers (cf. pp.18ff).

Mapping of the upper reaches of the glaciers and their contours proceeded very slowly. The *USGS Reconnaissance Map: The Alaska Railroad, Seward to Matanuska Coal Field, 1924* (Scale 1:200,000) was based on USGS surveys from 1904 to 1916. It was the first map to show the approximate extent of many glaciers, but it gives no contours and leaves the area of the Sargent Icefield blank.

The invention of the airplane brought a new tool to the study of glaciers — aerial photography, which provides information about the entire course of a glacier. The U.S. Navy took aerial photos of Valdez Glacier in 1934. Field took "snapshot" quality aerial photos in 1935. Bradford Washburn, the pioneer in aerial photography of Alaskan glaciers, made his first professional photographs of Columbia and Shoup Glaciers in the late thirties. During World War II, the Army made tri-met aerial photographs of Prince William Sound. This is the first complete photographic record of the icefields. However maps based on them did not show contours or give accurate measurements of the upper elevations of glaciers. The USGS maps (1:63,360 series) published during the 1950s are based on US Air Force photos from 1950 and 1957. These were the first maps to show contours on the glaciers and information on the icefields. Mapping of most of the glaciers has not been revised on subsequent updatings by the USGS, which makes the maps an invaluable source for studying changes.

A most important step was taken in 1960 when Austin Post made his first photographic reconnaissance of all the glaciers of western North America for the National Science Foundation through the University of Washington. The USGS hired him to continue the project on an annual basis in 1964. There is a good aerial photographic record for most glaciers from 1960 to 1987, because Post has continued to take photographs since his retirement in 1982. Post's aerial photographs enabled glaciologists for the first time to look at a number of glaciers as a whole, including their snowline, firn limit, drainage areas, and movement over a period of time.

Although glaciologists and glacial geologists observed glaciers modifying the landscape and recorded landscape features formed by earlier glaciers, they were often frustrated in their attempts to assign dates to this activity. Some scientists believed that glacial advances and retreats could provide invaluable clues to the long-term fluctuations in the climate and help scientists to assess the impact of industrial activity on the atmosphere. Interdisciplinary studies by glaciologists and botanists finally bridged this gap. During the '30s, Cooper and Viereck used tree-core samples as a means of establishing the most recent date by which an area was overridden by glaciers. Since the '50s, radiocarbon dating of organic material such as duff layers has helped to establish when plant life developed in ice-free areas following the retreat of the Pleistocene Glaciers. Radiocarbon dating of logs left by a retreating glacier gives important information on when the glacier overrode a forest. Heusser has used pollen samples to determine changes in vegetative communities that are indicative of warming and cooling periods in the climate. Finally, glacial geobotanists such as Crossen have turned to the study of lichens to pinpoint past fluctuations in the glaciers with more precision. The extent to which glaciers respond to climatic changes remains an open question, but interdisciplinary studies conducted in Prince William Sound and elsewhere around the world show a complex and intriguing picture.

Computers have focussed attention on developing models to predict glacial dynamics, ushering in a new era devoted to theoretical studies aimed at uncovering the physical laws which determine glacial flow and using these to predict the behavior of a glacier. Meier and others made an important contribution in this direction, (even though the model has been less than accurate) with their computer modelling of the predicted disintegration of Columbia Glacier (1980). Although the emphasis is now on developing computer models, the need for detailed data about the behavior of individual glaciers remains critical. Computer models designed to predict the behavior of glaciers require accurate data in both their development and verification.

Many questions about glaciers that puzzled early scientists have now been solved. However, probably the most important and most difficult unresolved problem is the relationship between a glacier and its bedrock basin. It is also the area where it is most difficult to collect the data essential to formulating and testing theories. Work by Kamb and others along these lines have combined observable data with computer modelling to explain the changes that occur in the relationship between a surging glacier and its bedrock before, during and after a surge. (Kamb 1985).

Chapter 9. RETREAT: Coastal Glaciers of Passage Canal and the Kenai Mountains

Location and Access:
Situated on the western side of Prince William Sound only 60 miles from Anchorage, Passage Canal marks the dividing line between the Chugach Mountains on its northern side and the Kenai Mountain Range on its southern side. It is one of the most frequently travelled fiords in Prince William Sound. The Port of Whittier, which is accessible only by boat or train, lies at the head of Passage Canal beneath Whittier Glacier.

Two icefields and a smaller snowfield feed most of the glaciers. The largest, the Sargent Icefield, stretches 37 mi. from northeast to southwest and 25 mi. from northwest to southeast, and drains into the Nellie Juan River, Kings Bay, Port Nellie Juan, Icy Bay, Port Bainbridge, Day Harbor, Johnstone Bay and Puget Bay. It is unique among icefields, because it has the lowest elevation accumulation area in North America outside of Queen Elizabeth Island in northern Canada. The Whittier Icefield stretches 37 miles northeast to southwest and 10 miles from northwest to southeast. It drains into the Snow River, Trail Creek, Placer River, Portage Creek, Passage Canal, Blackstone Bay, Nellie Juan River and Kings Bay. Its highest peak is located between Trail and Spencer Glaciers. Twelve of its glaciers terminate below 500 feet and two are tidal. A much smaller snowfield feeds glaciers along the northern side of Passage Canal, Pigot Bay, and Harriman Fiord.

The Kenai Mountains are much lower than the Chugach; the highest summit in the Sargent Icefield is a mere 6114 ft. and 6202 ft. in the Whittier Icefield. By contrast, in the nearby western and central Chugach Mountains, Mt. Marcus Baker reaches to 13,176 feet and Mt. Witherspoon to 12,021 feet. Four major fiords penetrate the Kenai Mountains from Prince William Sound: Passage Canal, Blackstone Bay, Port Nellie Juan with Kings Bay at its head, and Icy Bay with its tributary, Nassau Fiord.

Passage Canal: Pleistocene Glaciation of Passage Canal
Several glaciers flowed into the main Passage Canal Glacier during late Pleistocene times. Evidence of their former existence can still be seen. Along Passage Canal's southern shore a small, but nearly perfect, U-shaped valley bisects the ridge running between Whittier and Shotgun Cove. Shotgun Cove, itself, is a drowned valley hanging above Passage Canal. Along the northern side there are several cirques plus Billings, a valley glacier, and Seth, a cirque glacier. When the ancient Passage Canal Glacier underwent its retreat to the head of the fiord, Whittier, Portage and Learnard Glaciers remained.

Portage Glacier:

Location: Mountains between Passage Canal and Turnagain Arm
Access: By car: Turn off Seward Highway at Portage for the U.S. Forest Service viewing station and Portage Visitors Center. By foot: Portage Pass Trail from Whittier.
Type: Ice-calving valley glacier flowing from Whittier Icefield; calves into Portage Lake.
Length: 5.5 miles (Field 1975)
Area: 12.7 sq. miles (Field 1975)
Slope/aspect: North.
Status: Drastic retreat.
Photos: Fig-45.

Recent History:

Considerable controversy has occurred concerning the historic activity of Portage Glacier. Vancouver (June 1794) reports the route across the pass but does not mention that it involves crossing a glacier:

> The intermediate space [between Passage Canal and Turnagain Arm] was the isthmus so frequently alluded to before, on either side of which the country was composed of what appeared to him [Whidbey] to be lofty, barren, impassable mountains, enveloped in perpetual snow; but the isthmus itself was a valley of some breadth, which, though it contained elevated land, was free from snow, and appeared to be perfectly easy of access. (Vancouver 1798, 800-801)

As previously mentioned, Vancouver's journals are not as useful as could be wished, because he does not distinguish between snow, ice, and glaciers. It seems probable from this entry that Whidbey, on whose reports Vancouver based his journal entries, did not try to cross the pass. His job was to find the elusive Northwest Passage (or at least to prove its non-existence), not to check out trails. It is also possible that Portage's eastern lobe was behind the lip of the ridge and not visible to the Whidbey party below, that the area was still under snow in June, or that the pass, as is frequently the case, was overcast. Unfortunately, Tarr and Martin (1914) surmised from this passage that during Capt. Vancouver's time, Portage Glacier did not occupy Portage Pass. They argued that the route reported by Vancouver across Portage Pass must have been on land because the local Indians they met were loath to cross glaciers. From this, they concluded that Portage Glacier must have advanced rapidly since Vancouver's time, until reaching its maximum in the late 19th century.

Tarr and Martin's supposition that the sound's native peoples avoided glaciers as transportation corridors is countered by the evidence that Valdez Glacier was routinely used by the interior and coastal native peoples as a trading route. About 1850, a massacre of one of the trading parties on their way out Valdez Arm led to a shift to the Copper River route (Abercrombie 1900). The crossing of Valdez Glacier is much longer and more dangerous than the Portage Glacier crossing.

Perplexed by the problems surrounding Tarr and Martin's claim that Portage Glacier had made a major advance since Whidbey's visit, the Chugach National Forest Service sponsored glacial geologist Kristine Crossen to conduct investigations of the lichens and revegetation of the Portage Glacier area. Her work shows that the glacier occupied the pass during most of the nineteenth century (probably since 1810). She found no evidence supporting either a major 1890 readvance or an earlier, larger Portage Lake. At present, her research does not preclude an ice-free or partially ice-free pass during the 1700s. However, from what is known about the retreat and advance of iceberg calving glaciers, it is extremely unlikely that Portage Glacier could have rapidly pushed a depositional moraine across a former Portage Lake between 1794 and 1810 (cf. pp 19-21). According to Crossen's studies of tree-rings and lichens, in 1898 Portage Glacier lay about 3/4s of a mile from Passage Canal, had no lake on the western side, and was approximately 7 miles from Turnagain Arm.

Considerable changes have occurred at Portage Glacier since the turn of the century. In 1898, three tributary glaciers fed the glacier from the southern side and one or two from the northern side. During the intervening years, Shakespeare and Burns Glaciers separated from Portage, and the glacier retreated from the Portage Pass area. It still occupies a valley below Portage Pass on the west side from whence it flows into Portage Lake. In 1898, Portage Lake was not in existence. During the early 1960s it was a moderate sized lake, but as Portage Glacier retreats, Portage Lake grows.

Because Portage Glacier terminates in a lake, its behavior is more like that of tidewater glaciers than a land-terminating valley glacier. Portage Glacier's drastic retreat beginning in 1935 is similar to the retreat that started in 1984 at Columbia Glacier.

Exploration and Economic significance:

Eskimos and Indians, Russian fur trappers and Army explorers, early prospectors and the Alaska Railroad Commission, the U.S. Army and cross-country skiers all have traversed Portage Glacier as a route between Prince William Sound and Cook Inlet. From Vancouver's journals we first learn of the presence of a Russian-style cabin at the base of the pass. During the gold rush at the turn of the century, many miners crossed Portage Pass to reach the Sunrise and Iditerod Mining districts. In 1898, as part of its mission to help pioneers open up the frontier, the U.S. Army sent an exploratory team to find the best route from the coast to the interior. Of several reports written about explorations that year which mention the Portage Glacier route, W.C. Mendenall's is one of the best:

> The isthmus which connects Kenai Peninsula with the mainland is only about 12 or 13 miles broad from tidewater to tidewater and probably stands but little above sea level, but for 5 miles of this distance it is buried under a glacier, which flows from the high mountains of the peninsula to the south. This glacier at its highest point is about

The Glaciers of Prince William Sound, Alaska

Figs-45,45. *Top:* Portage Glacier is on the right with Burns Glacier on left — taken from Portage Pass (1971). *Bottom:* Portage Glacier is on the right with Burns Glacier just barely in view on left. Taken from near the same location as first photo but in 1983. The dark moraine next to the glacier is the old medial moraine. Photos by Nancy Simmerman.

1,000 feet above tide, and can be crossed in a few hours from the open waters of Portage Bay [Passage Canal] by prospectors or others who desire to reach Sunrise City or the headwaters of Cook Inlet before this body of water is open to navigation in the spring. For more than 100 years it has been used as a route, first by the Russian and Indian traders, and later by miners, who usually cross it without difficulty in the winter or early spring. In the summer the crevasses open, and it is but rarely used, especially since at that season the all-water route is so much easier and cheaper. (Mendenhall 1900)

In 1913, W. C. Guerin headed an investigative team searching for a railroad route to the Matanuska Valley coal fields. He landed in Passage Canal and crossed Portage Glacier to Turnagain Arm. Although the route was shorter and would have been less expensive to build than the route from Seward, Guerin rejected it. According to the *Valdez Daily News Miner* (1914) after the founding of Anchorage, an attempt was made to establish regular boat service from Valdez to the head of Passage Canal. Travellers spent the night at a roadhouse at the head of the canal, then crossed the pass. Another boat took them from the Kern roadhouse on Turnagain Arm to Anchorage.

The outbreak of World War II caused the construction of the Whittier Railroad Cut-off as an alternative to the Seward route should the Japanese attack Seward. Construction began in the fall of 1941. During the winters of 1941-1942 and 1942-1943, the U.S. Army maintained mail communication between Whittier and Anchorage by special mountaineering troops who crossed Portage Glacier regularly with crampons and ice axes. In 1943, the Whittier railroad tunnels and cutoff were completed. Since then, the glacier route has been virtually abandoned to occasional groups of cross-country skiers and mountaineers. Portage Glacier's considerable retreat in the last two decades has made the route increasingly difficult. It may now be impassible.

Routes to the Glacier:

Since the beginning of regular train service to Whittier in the 1960s, Portage Pass and Portage Glacier have been a very popular destination for day hikers. To reach Portage Pass follow the road between Whittier and the Army tank farm. Near the airport, a dirt sideroad crosses the railroad tracks. Take this road. When it branches, follow the right hand (western) branch which crosses a privately owned gravel pit before beginning the ascent to Portage Pass (750 ft.) just at the base of the avalanche slopes. At the summit, the panorama includes Burns, Shakespeare, and Portage Glaciers in the foreground, Portage Lake and Valley in the middle ground, Turnagain Arm, 12 miles away, and the Kenai Peninsula's mountains in the distance. One can stop here near the 1810 moraines and some glacially scoured tarns or continue a mile farther past Divide Lake to Portage Glacier. Beyond Divide Lake, the route becomes encumbered by patches of entangled alders — a refuge for sleeping black bears. But, for the watchful and patient hiker, the walk across a recently revegetated landscape dotted with clearings is well worthwhile.

Portage Glacier's ice-calving terminus jutting out into Portage Lake can be observed from numerous view-points. This is an excellent place to watch for ice-calving events and to observe such features as lateral and medial moraines, foliation patterns, seracs, crevasses, and, if one descends to the glacier, ice crystals. Care should be taken in approaching the glacier, as the seracs are unstable and may fall at any time.

Whittier Glacier:

Location: Above the town of Whittier
Access: Cross-country hike from Whittier.

Type: Hanging glacier flowing from Whittier Icefield.
Slope/aspect: North.
Status: Retreating.

Whittier Glacier is now barely visible behind the town, but in 1910, Tarr and Martin found it descending part way down the cliffs to terminate several hundred feet above sea level about 3/4s of a mile from Passage Canal. In 1941, the glacier generated enough ice avalanches that the Army classified Whittier as one of the few towns in Alaska with an ice avalanche hazard. Periodically, writers reporting on Whittier use these outdated materials without noting the changes. Total recession between 1910 and 1971 was 3940

Fig-46. Whittier Glacier (1981) has retreated throughout this century. When the Army built Whittier in the early 1940s, ice avalanches posed a hazard. Photo by N. Simmerman.

ft. horizontally and 575-740 ft. vertically (Field 1975). It is still retreating. The Prince William Sound Recreation Association plans to construct a trail to the glacier.

Billings Glacier:

Location: North side of Passage Canal
Distance from Whittier: 3 miles
Access: Boat to outwash stream. No trail

Type: Valley glacier flowing from small icefield
Length: 5.5 miles (Field 1975)
Area: 4.6 sq. miles (Field 1975)
Slope/aspect: SSW
Status: Slow retreat

Billings Glacier, the largest glacier in Passage Canal, is fed by the same icefield that feeds Pigot and Harriman Glaciers; and, like all other glaciers in Passage Canal, it has been retreating throughout this century. Tarr and Martin (1910) estimated the terminus to be about 1 to 1 1/2 miles from the fiord. Field (1975) calculated that between 1910 and 1971, Billings Glacier retreated 1 mile and possibly more. Barren areas around Billings indicate it is still both shrinking and retreating.

In the last decade, a patch of light colored granitic rock has appeared near the eastern side of the glacier's terminus. This is part of a granitic batholith that probably runs beneath the ridge between Billings and Seth Glaciers. An impressive, ice-carved gorge separates the low-lying bedrock hill in front of Billings from this ridge.

Fig-47. Billings Glacier (1984) has retreated throughout this century. Dead trees in the foreground are from the land subsiding during the 1964 Earthquake. Photo by author.

Seth Glacier:

Location: Northern shore of Passage Canal
Distance from Whittier: 6 miles

Access: Boat and cross-country hiking. No trail.
Type: Cirque
Length: 1 1/2 miles
Slope/aspect: Southwesterly
Status: Retreating
Photograph: Fig-2.

Seth Glacier lies in a cirque at the head of Poe Valley. A waterfall flows from the small ice-tongue extending from the glacier at the head of the valley. At the base of the cliffs below the ice-tongue, almost at sea level, there is a permanent snow and ice patch created by ice and snow avalanches.

Seth has been retreating throughout this century. In 1910, Tarr and Martin reported that it was about 1 mile from the water. Field found it shrinking and retreating in 1935. By 1950, Seth Glacier was 3 miles from Poe Bay. Unweathered, barren rock on the ridges show that the glacier and icefield are continuing to shrink.

The Portage Bay Mine (gold and silver) belonging to the Veitti family was located on the cliff beneath the ice tongue. Earlier in this century, when the glacier stretched almost to Passage Canal, prospectors freighted their goods and supplies up Seth Glacier. By 1966, the glacier had shrunk so much that the mine was almost inaccessible.

Tebenkof Glacier:

Location: Mouth of Blackstone Bay
Distance from Whittier: approx. 10 miles.

Access: Boat & hike.
Type: Valley glacier flowing from icefield
Length: 8 miles (Field 1975)
Area: 10.8 sq. miles (Field 1975)
Slope/aspect: North/northeasterly
Status: Slow retreat

Tebenkof, named for Capt. M.D. Teben'kof the Governor of Russian America from 1845 to 1850, is the only glacier in Prince William Sound to split into three streams: the main stream flows north towards Passage Canal; a second stream flows east towards Cochrane Bay, and a third flows west towards Blackstone Bay.

Tarr and Martin (1914) found the mainstream within 1/2 mile of tideline, but it had been retreating slowly; the 1964 terminus stood 3620 ft. back from the 1885 moraine. The prominent knoll and gorge on the west side was ice-covered in 1910. By 1935, the glacier had retreated off the knoll, leaving in its wake remains of a buried forest (Field 1975). The gorge was probably formed by a glacial stream finding a weaker zone of bedrock and making a permanent channel there under the ice.

Tebenkof Glacier illustrates how a valley glacier's terminus spreads out into a bulb-shaped fan after leaving the valley and flowing out onto a plain.

Fig-48. Looking south into Blackstone Bay with Willard Island in the right foreground. Glaciers from left to right are Lawrence, Marquette, Beloit (ice-calving), Blackstone (ice calving), Northland, and Concordia. The point between Beloit and Blackstone Glaciers was just barely ice-free when Grant and Higgins made the first map of the area on July 5, 1909. Cooper (1942) found a tree with 417 annual rings on it here, so recession for four centuries had been very slow. Photo by Nancy Simmerman.

Blackstone Glacier.

Location: Head of Blackstone Bay
Distance from Whittier: Approx. 20 miles
Type: Ice-calving valley glacier from

Whittier Icefield
Length: 6.8 miles (Field 1975)
Area: 31 sq. miles (Field 1975)
Slope/aspect: Northeasterly
Status: Slow retreat
Photo: Fig-48. Map: Fig-49.

At various times during the Pleistocene and neoglacial periods, the group of glaciers at the head of Blackstone Bay joined and advanced down the fiord. A prominent moraine crosses Blackstone Bay from Willard Island. Post reports that the fiord north of the moraine was occupied only by the Pleistocene glaciers, because the bottom here is covered with deep sedimentation, while the bottom south of it has little sedimentation, indicating that more recent glaciation has scraped off the earlier layers of sediments. Post dates the Willard Island moraine to a neoglacial advance (Post 1980c). Radiocarbon dating of peat samples found near the moraine show them to be 580± 55 years old (Heusser of peat samples found near the moraine show them to be 580± 55 years old (Heusser 1978). Post concludes that "The glaciers thus had retreated to the head of the bay before 1400 and suggest drastic retreat from the terminal moraines probably occurred around 1350, if the rate of retreat of present-day retreating tidal glaciers is used as a guide (Post 1980c)." Subsequently, Blackstone and Beloit Glaciers established new terminal moraines south of the Willard Island moraine. These can be easily spotted with a depth sounder. It is not known when Blackstone retreated off its second moraine to its present position, but Post suggests that the retreat has been in the last 100 years. Beloit remained on its moraine until around 1950, then retreated.

Blackstone Glacier looms in the southwest corner of the bay where it descends several hundred feet in the last quarter of a mile. In the early 1980s, its eastern front retreated to several feet above hightide line. In 1987, it was again tidal. Sometimes dramatic calving events occur when large seracs plunge down its ice-cliff. Waves large enough to damage small boats and kayaks have been reported.

In 1987, the Wetco Company of Anchorage collected icebergs from Blackstone Bay. The ice was then shipped to Tokyo where 2.2 lb. bags retailed for about $6.

Other Glaciers:

Eight smaller glaciers descend the precipitous walls of Blackstone Bay. They have all been shrinking and retreating slowly throughout this century, except for Lawrence Glacier where in 1966 Field found push moraines in front of it indicating small advances.

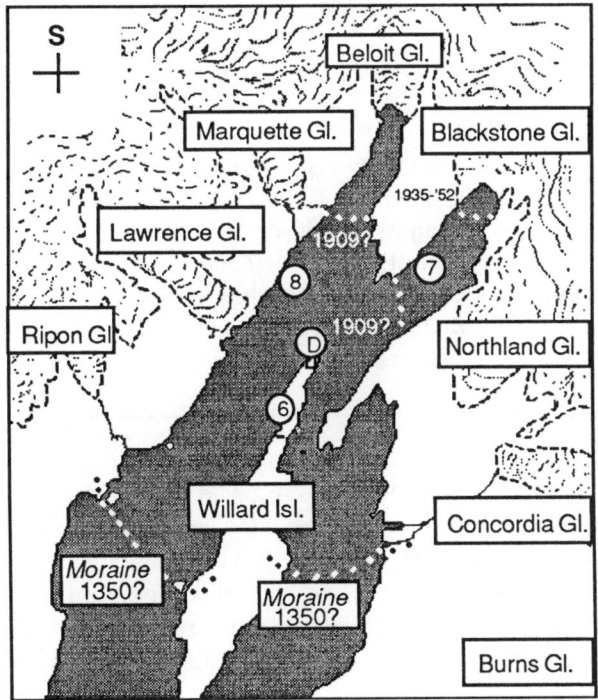

Fig-49. Blackstone Bay

Adapted from Post, *Preliminary Bathymetry of Blackstone Bay and Neoglacial changes of Blackstone Glaciers, Alaska* (1980).

Photographic Stations:

Station D on the southern tip of Willard Island was established by Grant and Higgins in 1909.

Stations 6 atop a knoll near the southern end of Willard Island, 7 on a rock, and 8 on a point were established by Field in 1937.

An Observer's Guide to

Port Nellie Juan

Glacial Features and Bathymetry of Port Nellie Juan:

The bathymetry of Port Nellie Juan is typical of fiords. At its mouth, a sill with less than 1200 ft. of water over it rises to separate the deeper waters of the fiord from Perry Passage. A second sill of less than 900 ft. stretches from Deep Water Bay to the northern shore. A third sill, covered by less than 600 ft., arches from the western side of Greystone Bay to Shady Cove. Between each sill are deep basins (Jones 1974).

Although obviously a glacially scoured fiord, Port Nellie Juan seems to lack the overly steepened sideslopes of the typical glacially carved, U-shaped fiord. However, a look at NOAA chart No. 16700 shows that the U-shaped configuration is preserved underwater. In many places, depths plunge suddenly to over 600 and 1200 ft. within a few hundred yards of the shoreline. The deepest sounding is 2052 ft., making Port Nellie Juan the sound's deepest auxiliary fiord.

The deepest sounding in Prince William Sound is 2460 ft. off Lone Island. This is the deepest sounding along the western North Pacific coast. Port Nellie Juan's 2052 ft. is the second deepest place north of Humphrey Channel (2250 ft.) in Canada.

Ultramarine Glacier:

Location: Head of Blue Fiord.
Distance from Whittier: Approx. 41 miles
Access: Boat to head of Blue Fiord; cross-country hike.

Type: Valley glacier flowing from Sargent Icefield
Length: 5.6 miles (Field 1975)
Area: 12 sq. miles (Field 1975)
Slope/aspect: Northeasterly
Status: Slow retreat
Photo: Fig-12.

Ultramarine Glacier terminates beyond an outwash plain covered with dense alders and several terminal moraines a little less than a mile from tideline at the head of Blue Fiord. Applegate's map showed Ultramarine as tidal in 1887, but Grant and Higgins (1913), on the basis of vegetative studies, doubted that either the eastern or western front had been tidal recently. Their surmise was later supported by more detailed work by Viereck (1935) which established that Ultramarine could not have been tidal in 1887. Ultramarine Glacier is rarely visited and deserves additional study.

Nellie Juan Glacier:

Location: Head of Derickson Bay
Distance from Whittier: Approx. 40 miles
Access: Boat; easy cross-country walk to terminus
Type: Ice-calving valley glacier flowing from Sargent Icefield
Area: 19 sq. miles (Post 1986 unpublished data)
Firn limit: 1300 ft (Post 1986 unpublished data)
AAR: 91% (Post 1986 unpublished data)
Aspect: East/northeasterly
Status: Drastic retreat
Photos: Figs-4, 6, 12, 50-53.

Nellie Juan Glacier, at the head of Derickson Bay, is the scenic focal point of Port Nellie Juan. Only two ice-calving glaciers in the Sound (Nellie Juan and Shoup) have anchorages within a reasonable day's hike, adequate space and water for camping, and extensive, recently glaciated areas to explore. Nellie Juan has the most secure anchorage, the best hiking, and the most easily observed glacial and glacially formed landscape features.

Recent History:

Vegetative studies by Viereck show that Nellie Juan's position in 1880 was its most advanced in at least two centuries. Radiocarbon dating of logs indicates Nellie Juan overrode a 12th to 13th century forest. Glacial maximum was reached between 1860 and 1880. Like most glaciers in the area, Nellie Juan has undergone a major retreat in this century. Fortunately, its activities have been documented. The earliest account is by Grant and Higgins (1913) who reported:

> ... the (terminus) rests on a gravel beach, most of which is covered by high tide; near the center of the front the ice is bathed even at low tide water. On each side of the lower part of the glacier is a distinct bare zone of smoothed granite, 100-500 feet wide, which ends abruptly at the edge of a forested tract. This zone is prominently developed on a granite knob, almost an island, at the west side of the glacial front. Crossing the top of this knob is a small moraine from 1 to 10 feet high and 5 to 30 feet wide. This moraine contains fragments of dead wood, and just north of it is an area of scattered trees some of which are a foot in diameter. . . . most of the vegetation disappears halfway to the ice front (p. 46).

The small granite knob described above is now a tree-covered knoll in the middle of the former outwash plain on the west side of Derickson Bay. By 1986, Nellie Juan Glacier had retreated over 3 miles from here and was no longer visible from the photographic station (marked by a cairn) that Grant and Higgins established.

In the years following Grant and Higgins' visit, Nellie Juan did not become a valley glacier, like Ultramarine, but slowly moved off its moraine into a deep lagoon. As is typical for tidewater glaciers, Nellie Juan first developed embayments; with the decrease

Fig-50. In 1935, Nellie Juan Glacier still rested on the moraine separating modern Nellie Juan Lagoon from Derickson Bay. Barren areas on the eastern side show the glacier is retreating. It is no longer visible from this photo station. Uplift by the 1964 Earthquake raised the moraine. Photo by Field, 9/12/35 from World Data Center. Photo damaged.

in pressure at the embayments, more ice flowed down from the icefields. This drew its icefields down further resulting in shrinkage in the height. In 1935, Field reported that the west side was in deep water, while the east side shoved forward a prominent push moraine. The 1935 push moraine is still visible as a small ridge running along the top of the larger moraine. Alders and herbaceous vegetation have become well-established on it.

By 1964, Nellie Juan had retreated back to the prominent point that hikers now cross on their way to the glacier before skirting the basin of a formerly ice-dammed lake. At several places along the shoreline, Field noted large stranded icebergs. Piles of gravel (kames) and depressed areas (small kettles) still mark their location.

When Field returned again in 1974, Nellie Juan had retreated another half-mile where it was just beginning to enter the narrows. Only a small portion of the rock bluff that now must be climbed to reach the glacier showed in 1974. During the late seventies and early eighties, Nellie Juan steadily continued to retreat; the retreat was slowed by the constricted channel between the two ridges and possibly by a shallower area in the middle of the channel. In August of 1986, we observed that Nellie Juan had retreated behind the cliff overlooking the narrows into what appears to be the start of a wide basin. According to Post's aerial photographs, between 1981 and 1986 Nellie Juan retreated approximately 500 ft. on its south side, 1000 ft. in the middle, and 1000 ft. along its northern margin. As more of its face is now exposed to salt water, calving activity has increased dramatically as evidenced by numerous icebergs, considerably larger than before, floating

The Glaciers of Prince William Sound, Alaska

Fig-51. Nellie Juan Glacier, Aug. 12, 1961. Photo by Post, USGS.

Fig-52. Nellie Juan Glacier, Sept. 10, 1972. Photo by Post, USGS.

Fig-53. Nellie Juan Glacier, Sept. 13, 1986. Photo by Austin Post. USGS. This series of photos shows changes in the terminus over the past 25 years. Note also the changes in the lakes and drainage patterns. The light colored rock to the right of Nellie Juan Glacier has recently become ice-free.

and aground in the lagoon. Glaciologists do not know whether Nellie Juan's continued retreat will open up another inner lagoon or if it will finally retreat onto land.

As Nellie Juan retreated, glaciologists had an opportunity to record a number of changes in the landscape. To the south and southeast of Nellie Juan lie several interesting features, including another former outwash plain, moraines and a kame, and the strandlines of a former glacier-dammed lake. This lake formed around 1935 as Nellie Juan retreated and drained sometime before 1957. It is still shown on the USGS contour map. Hikers to the glacier can see the strandlines from the lake. Between the old lake bed and the narrows lies a side valley filled with morainal piles and several small lakes. An ice tongue that was formerly part of Nellie Juan Glacier has retreated up this side valley.

To the west of Nellie Juan between the former outwash plain and its present terminus lies a barren hanging valley with a large waterfall. In 1935, ice flowed from the main glacier into this valley, which may have at one time had a tributary glacier feeding Nellie Juan. The Army's 1941 aerial photographs show that by then the ice-tongue had almost disappeared, leaving a glacier-dammed lake. Between 1941 and 1961, the ice-tongue continued to recede and the glacier-dammed lake split into two lakes. By 1961, the ice-tongue ceased to dam the lakes; their level and output became determined by their rock basins.

As Nellie Juan retreated, marine life moved into the lagoon, which is connected to Derickson Bay by a boulder-strewn, tidal stream that is only navigable by shallow draft boats at high water. By the late seventies, pink salmon were seen spawning in the intertidal zone at the base of the first two waterfalls flowing from lakes on the southeastern side. In 1984, we observed a few salmon trying to spawn in the stream that flows across what was formerly the bed of the ice-dammed lake. Bald eagles, glaucous-winged gulls, and mew gulls congregate at the mouths of the streams during the salmon runs. Numerous birds, including rock ptarmigan, spotted sandpipers, semipalmated plovers, black oystercatchers, mew gulls, arctic terns, and buffleheads, nest in the lakes, cliffs and morainal areas. Hikers should exercise care not to disturb the birds during nesting season (May to August). Within a few years of the glacier's retreat from the area, black-legged kittiwakes established a small nesting colony on the cliffs above the narrows. At times, a hundred or more harbor seals haul-out on icebergs trapped in the lagoon. It is possible, although unconfirmed, that harbor seals pup on the icebergs as they do near other tidewater glacier fronts. In June, mothers with young are seen on the icebergs.

Routes to the Glacier:

From the small boat anchorage, it is an open-country walk over the ridge to a viewpoint of Nellie Juan Glacier calving into the lagoon. From here there are two routes. One can descend the broken rock cliff to the beach and walk to the glacier. This involves crossing a large stream, which even at low tide is often knee deep. Or, one can cross the stream via some large boulders about 1/8 mile above the waterfall, skirt the outwash valley for a couple of hundred yards, then follow a small creek down a cut that crosses the 1885 moraine to the beach. Along this route, one passes a lateral moraine and kame and can observe the spread of vegatation and small animals into the recently deglaciated area. Once on the beach, there are several more streams to cross. At the headland, the route goes over the point and down to the bed of the former glacier-dammed lake, which may be followed until reaching the rock cliffs near the narrows. From here, one can pick a route over the point to the glacier. Most of this area was covered by the glacier in 1977. Because the lagoon is tidal, it is best to start when the tide is well into its ebb and plan to return before it reaches the middle of the flood.

At hightide, one can also take skiffs up the tidal river to the lagoon in front of Nellie Juan Glacier. Currents are strong. And, at ebb, icebergs race down the river. The river, encumbered by many boulders, is not recommended at the lower stages of the tide. Many prefer to take their inflatables and kayaks to the moraine and carry them across to the lagoon. Care should be taken when leaving skiffs and kayaks along the shore to have them well above the tideline; major calving events can generate waves several feet high that have been known to swamp boats or wash them out into the lagoon.

An Observer's Guide to

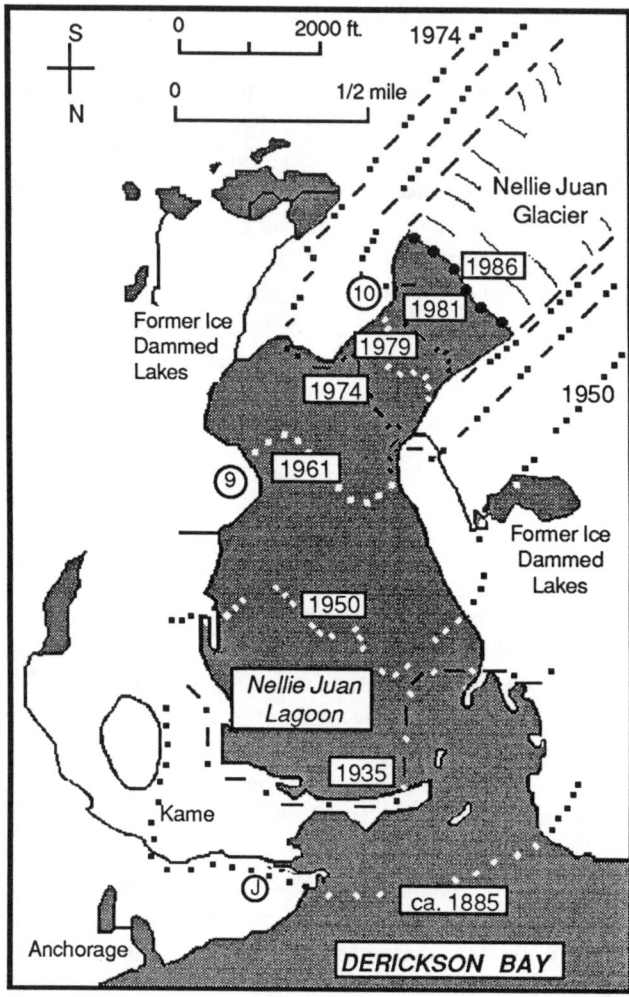

Fig-54. Terminus of Nellie Juan Glacier: 1885-1986

Sources:

Field, *A.G.S. Glacier Studies*, Map No. 64-4-G7. Based on Field 1935 trip. USGS Seward, B-4, AK. 1:63-360 series (1950 aerial photographs.) 1961 Field, Brigham Young Univ. Survey.

Austin Post, USGS aerial photographs: 1974, 1979, 1981, 1986.

Photographic Stations:

Photographic Station J is atop a prominent knoll above the camping area on the spit just before crossing the first stream. It affords good panoramic views of the moraine, former outwash valleys (east and west), Nellie Juan Lagoon, and some long distance shots of the glacier's terminus. In 1966, Field established Photographic Station 9 on top of the prominent point that must be crossed by hikers on their way to the glacier. Since Nellie Juan's retreat has rendered earlier stations useless, this is the best historic station to use. Station 10 (Lethcoe) has been established on the bluff above the narrows. This is a good spot from which to photograph the glacier as it retreats into the basin.

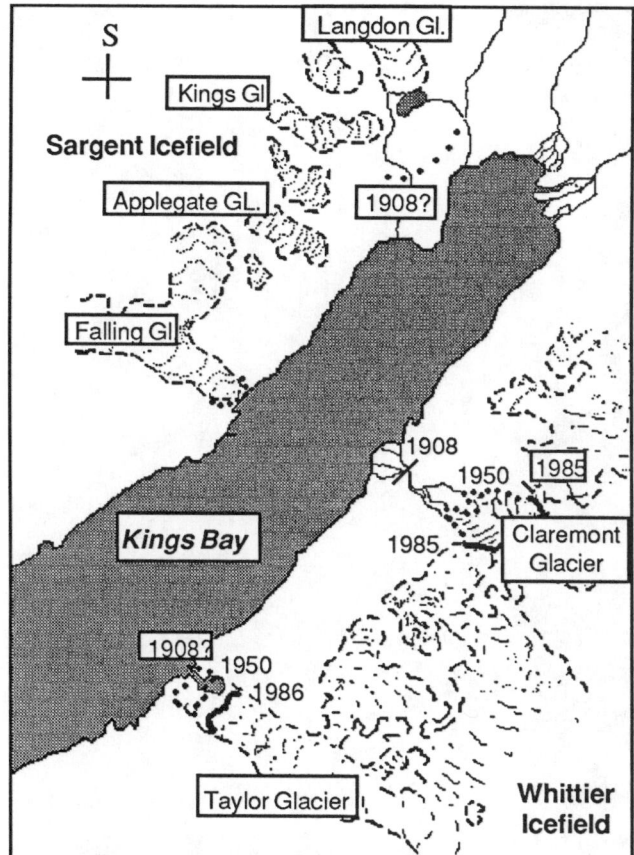

Fig-55. Kings Bay

Sources: USGS Seward B-5, C-5, Alaska, 1:63,360, 1952 (based on 1950 aerial photography).

Field, *Glacier Termini in Kings Bay: Prince William Sound Alaska, 1964.*

Post, 1986 aerial photograph

Taylor Glacier:

Location: Western side of Kings Bay, Port Nellie Juan
Distance from Whittier: Approx. 55 miles
Access: Boat and cross-country hike

Type: Valley glacier from Whittier Icefield
Length: 5 miles (Field 1975)
Area: 9 sq. miles (Field 1975)
Slope/aspect: East southeasterly
Status: Retreating
Photo: Fig-56
Map: Fig-55

Taylor Glacier has been steadily retreating since first observed by Grant and Higgins in 1909. Field (1975) has inferred from vegetative studies that Taylor reached its maximum sometime between 1867-1872, then began to recede sometime between 1872 and 1908. In 1909, Grant and Higgins reported Taylor Glacier to be partially tidal. It remained partially tidal until sometime between 1935 and 1941. A small terminal lake formed

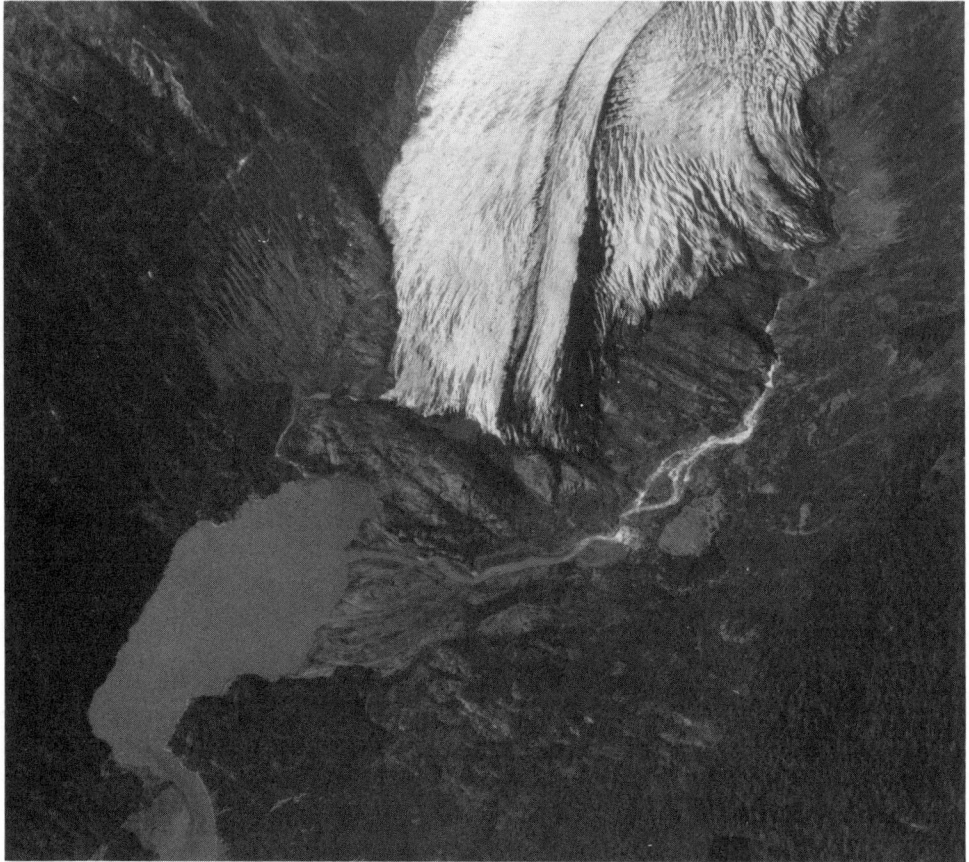

Fig-56. Terminus of Taylor Glacier with outwash stream and lake. Note medial moraines and longitudinal crevasses. Photo by Post, USGS, Sept. 10, 1980.

between 1941 and 1950. In 1950, the lake was 165 ft. long; in 1961, it was 1000 ft. long. Field estimates that the total retreat from 1870 to 1961 was probably about 3525 feet. During the 1964 Earthquake, the land subsided 1.2 feet, turning the lake into a tidal lagoon. Between Field's 1964 trip and his return in 1974, Taylor retreated on to dry land.

Photographic Stations:
Field offers the following suggestions for photographers. Taylor Glacier has "retreated behind a ridge which in 1950 was at the terminus. This ridge is now covered with dense alders and our stations on it would by now be difficult to locate. My suggestion would be to walk up the outwash plain on the east side of the creek to near the middle of the ridge. Here there is a small draw which used to give fairly good access to the inner side of the ridge from where you should be able to see the terminus. Rather than hunt for one of the stations, I suggest simply taking a series of pictures from the glacier side of the ridge

forming a panorama from the southwest side of the valley northward along the terminus to the northeast side of the valley. Be sure to include the features in front of both ends of the terminus down as far as the ridge." (Letter to author, July 1, 1986).

Until Field's visit in 1935, no one had photographed the terminus of Taylor Glacier. He rephotographed the glacier in 1961, 1964, 1966 and 1974. Since then, no land-based photographic studies have been made.

Claremont Glaciers:

Location: Western shore of Kings Bay, Port Nellie Juan
Distance from Whittier: Approx. 57 miles
Access: Boat and cross-country hiking.

Type: Valley glaciers from cirques
Slope/aspect: Southeasterly
Status: Retreat
Photo: Fig-57
Map: Fig-55

Claremont Glacier, first described by Field using the photos and maps of earlier visitors, is made up of two branches, which early photos show joined immediately above the terminus as depicted on NOAA chart 16700 (1986). In 1961, the two branches became completely separated. Field estimated that the total recession between 1908 and 1966 was 4625 feet for the north branch and 7575 feet for the west branch (Field 1975). The terminus of the northern branch sports an interesting medial rock debris line. Aerial photographs by Post suggest that its source might be a rock avalanche zone.

Field (1975) recommended the rock ridge in front of Falling Glacier as a good photographic station for Claremont Glacier. In 1986, the western terminus was scarcely visible from this station.

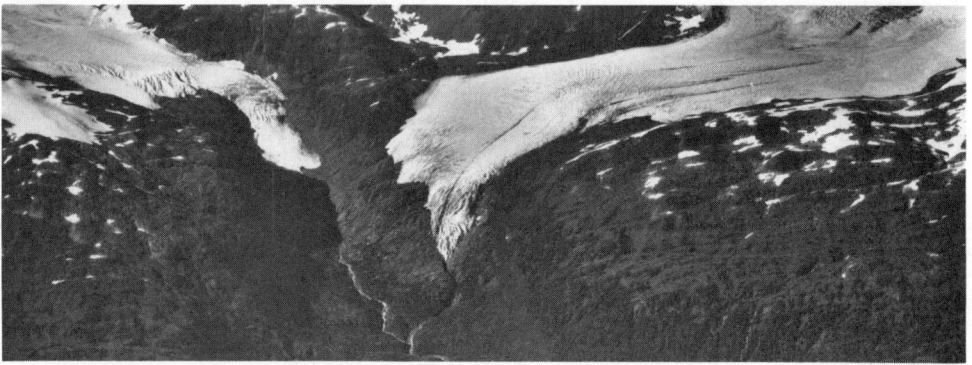

Fig-57. Claremont Glacier. Sept. 3, 1966. Photo by Austin Post. USGS.

An Observer's Guide to

Falling Glacier:

Location: East side of Kings Bay
Distance from Whittier: Approx. 58 miles
Access: Boat and cross-country hiking; no trail.

Type: Valley glacier from Sargent Icefield
Length: 6.8 miles (Field 1975)
Area: 13.5 sq. miles (Field 1975)
Slope/aspect: Northwesterly
Status: Slow retreat
Photo: Fig-58
Map: Fig-55

Approached from the water, Falling Glacier comes suddenly into view through a narrow gorge that bisects a bedrock barrier, formerly overridden by the glacier. From the trimline on this bedrock ledge in 1908, Grant and Higgins estimated that at its most recent maximum reached sometime near the end of the last century, Falling Glacier extended to the 1200 foot level. From here it fell abruptly over the bedrock ridge to the sea. This must have been a truly remarkable sight, which no westerner ever saw. In 1986, the trimline between the old forest with its standing dead trees and the younger vegetation was still evident on either side. Following its significant retreat around the turn of the century, Falling Glacier remained remarkably stable from 1908-1935. Falling Glacier continued to be marginally tidal until about 1961, and as late as 1969 it still occupied the gorge. In 1986, we found it terminating a short distance back from the gorge.

In addition to the impressive gorge with its somnambulent glacier lurking in the depths behind, visitors will notice many distinctive light colored, granitic erratics on the sedimentary beaches, a prominent lateral moraine along the glacier's northern side, dark cone-shaped piles of rock debris from avalanches along the glacier's margin, and strong longitudinal crevasses in its terminus.

Fig-58. Near the end of the last century, Falling Glacier flowed over the 1200 ft. bedrock ridge before cascading abruptly to the water. Photo by author. Aug. 1985.

The Glaciers of Prince William Sound, Alaska

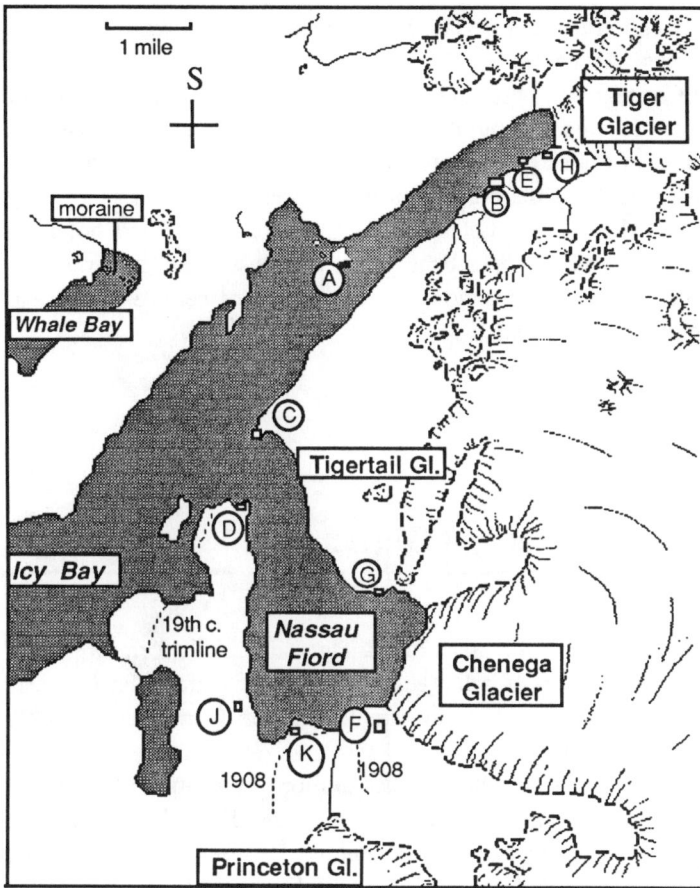

Fig-59. Icy Bay and Nassau Fiord.

Sources: USGS Seward A-4 and B-4, Alaska, 1:63-360, 1952 (based on 1950 aerial photography.)

Field, *Glacier Termini in Icy Bay, Prince William Sound, Alaska, 1957.* (map)

Field, 1986 revision of *Glacier Termini of Icy Bay* showing positions of photographic stations (personal communication).

Station F is the best photo station for Chenega Glacier (cf. Fig-63.) Stations A, B, and C were first used by Grant and Higgins. Stations G was used by Perkins in 1909. The rest were established by Field on AGS trips.

Icy Bay:

Icy Bay, in the southwestern corner of Prince William Sound, lies about 70 miles from both Whittier and Seward. A small arm, Nassau Fiord, penetrates the mountains to the northwest. Herein, lie three glaciers: Tigertail, Chenega, and Princeton. Icy Bay itself terminates at the face of Tiger Glacier.

Glaciers in this area all flow from the Sargent Icefield. Recent photographs of the termini of glaciers in Icy Bay and Nassau Fiord are very limited. This is an area where frequent visitors could make a contribution to the study of these glaciers by re-occupying established photographic stations.

Vancouver's journals mention ice in Icy Bay, but Applegate (1887) was the first to indicate a glacier on his map. According to Grant and Higgins (1913), "The Indians living at the settlement of Chenega have a tradition that the Chenega Glacier reached to the mouth of Icy Bay about 100 years ago, but the growth of the forest about the bay and even well up past the mouth of Nassau Fiord precludes this idea. The tradition would refer more reasonably to the mouth of the northern arm [Nassau Fiord] than to Icy Bay itself." (Grant and Higgins 1913, p. 49). Later studies by glaciologists and botanists supported Grant and Higgins' contention establishing that the most recent maximum of Chenega and Princeton Glaciers was at the mouth of Nassau Fiord between 1787 and 1898 and not at the mouth of Icy Bay (Field 1975).

Tigertail, Princeton and Chenega Glaciers:

On a narrow gorge on the southern side of Nassau Fiord, Tigertail Glacier forms a slender ice-finger extending down over the lip of a hanging valley in an icefall that formerly reached tidewater. Probably as a result of both morainal filling of its inlet and possibly some shrinking, it is now no longer tidal.

Princeton Glacier lies on the northwest side of Nassau Fiord. When first observed by George Perkins' party in 1909, its western ice-cliff was still tidal. By 1910, it ceased to be tidal and has continued its retreat throughout this century. Between 1908 (Grant and Higgins' map) and the 1950 aerial survey on which the USGS Seward B-4, 1952 map is based, Princeton retreated a little over a mile (Field, personal communication 11/26/86). Field (1975) noted that Princeton's icefields are relatively low and predicted that the glacier was approaching "relic" status. In the recently exposed area in front of Princeton lies a small lake which glaciologists found interesting because of the speed with which salmon drifted from their natal streams into this newly formed spawning habitat.

Chenega Glacier's tall ice-cliffs which are over 100 ft. high dominate Nassau Fiord. It descends 500-600 ft. in the last 1/4 mile with severe crevassing and considerable activity at its face. Since 1908, Chenega has been relatively stable despite its tremendous ablation rate through calving. With an AAR of 94% (Post, unpublished materials), however, this is hardly surprising. Chenega Glacier is 13.6 miles long and covers 143 sq. miles (Field 1975).

Field suggests that one explanation for Chenega's stability may be the bathymetry of Nassau Fiord. "The terminus may rest at the edge of an area where its bed slopes downward, so that any tendency to advance would be matched by increasing ice berg

Fig-60. Chenega Glacier from AGS Sta. F at about 500 ft. elevation on ridge north of terminus. Photo by Marion T. Millett for AGS, Aug. 16, 1957. No. AGS-IGY-MN-57-SG34.

discharge. This would tend to keep the terminus in about the same position over these past six decades of observations (Field personal communication 11/26/86)." Black-legged kittiwakes nest on cliffs near the terminus of Chenega Glacier.

Tiger Glacier:

Tiger Glacier lies curled at the head of Icy Bay. According to Field (1975), it is 6.8 miles long and covers 20 square miles. Tiger's vertical, ice calving cliff is less than 1/2 mile wide. From 1910 to 1957 Tiger advanced. It retreated an estimated 500 ft. between 1957 and 1966. Field reports that a conspicuous ledge appeared in the middle of the terminus. Tiger appears to be retreating slowly. Because icebergs and winter pan ice often clog Icy Bay, this area is rarely visited. There is a black-legged kittiwake rookery near the terminus of the glacier.

Chapter 10. Of Gentlemen and Glaciers: Port Wells, Barry Arm, and Harriman Fiord

Port Wells is the westernmost large fiord penetrating into the Chugach Mountains. It is also the most complex in structure. Thirteen miles up the fiord, Port Wells splits into two arms: to the west is 5 mile long Barry Arm which becomes 10 mile long Harriman Fiord and Surprise Inlet; while to the east lies College Fiord which has Yale Arm as its tributary (cf. next section).

An interesting meteorological phenomenon is often observed by boaters crossing the mouth of Port Wells. Barry and Cascade Glaciers may appear to have suddenly advanced three miles to a position in front of Point Doran. The apparent illusion is really an inferior mirage with towering which occurs when the temperature is warmer near the water and decreases with height. The resulting temperature profile has curvature which displaces the image downward from the object. Towering, which often accompanies inferior mirages, is found when both the temperature gradient and temperature increase with height, causing the image to be magnified.

Temperature profiles that give rise to towering inferior mirages occur quite frequently over enclosed bodies of water. In the vicinity of glaciers, a local high pressure area develops over the glacier relative to an area of low pressure over the adjacent warmer water. Cold air from the glacier flows down over the warm water and is heated from below. The resulting difference in temperature and density between the layer of warm air over the water and the cold air immediately above it causes the inferior mirages, making Barry and Cascade Glaciers look much larger and closer than they actually are.

Pigot and Bettles Glaciers:

Pigot Glacier lies at the head of Pigot Bay and shares its snowfield with Seth, Billings and Harriman Glacier. Applegate in 1887 was the first explorer to map Pigot Glacier.

Pigot Glacier's most distinctive features are the many snow and rock avalanches which descend on to the glacier from the northern wall above its middle branch. Photographs by Post show one nearly 2 miles long avalanche. Rock debris, originating primarily in the middle branch, covers the terminus. Photographs taken in 1964 following the earthquake showed numerous large slides covering the glacier with extensive debris. Since rock debris shades the ice from the sun's rays, Field predicted that Pigot's rate of recession would likely diminish or that it might even begin to advance temporarily as a result of reduced melting (1975).

Bettles Glacier shares its small icefield with Dirty and Harrima[...] north and Pigot Glacier to the south. It is one of the most enjoyed sce[...] sound as it terminates in icefalls about 1 1/4 miles from the head of B[...] ular anchorage. The thunder of ice-avalanches tumbling from its ice-cli[...], particularly in the vicinity of the major waterfall, can occasionally be heard from boats in Bettles Bay. Between 1910 (Tarr and Martin) and 1971 (Field 1975), Bettles Glacier retreated a little over a mile. Barren areas around the ice-falls suggest the glacier is shrinking, and photographs taken by the author over the last decade show that it is still retreating.

Barry, Cascade, Coxe Glaciers:

Location: Head of Barry Arm
Distance from Whittier: 34 miles
Access: Boat
Type: All three are ice-calving tidewater glaciers

Length of Barry: 15 mi. ; Cascade: 5.6 mi.; Coxe 6.8 mi. (Field 1975)
Area of Barry: 29 sq. mi.; Cascade: 5.8 sq. mi.; Coxe: 7.3 sq. mi. (Field 1975)
Status: Since 1979 Barry has retreated; Cascade has advanced; and Coxe is advancing.
Photos: Figs-31, 61-65.
Maps: Figs-41, 66.

Recent History:

When first seen by Whidbey in 1794, Barry, Coxe and Cascade were united and filled Barry Arm, extending almost all the way to Pt. Doran. Vancouver's map, based on reports by Whidbey, shows Barry Arm ending at Pt. Doran. When Applegate in 1887 and Lt. Glenn in 1898 visited Barry Arm, Barry Glacier still extended almost to Pt. Doran, calving ice and obstructing the view of the fiord beyond.

Glenn, who named Barry Glacier after Col. Thomas H. Barry, assistant adjutant general, U.S.A., called Barry "One of the most formidable as well as the most interesting of any (glacier) we have seen (Glenn 1899, p. 19)." He reported that Barry Glacier's calving events were so noisy that they could be heard at Yale Glacier, 15 miles away, and seemed to shake the mountains with their thunder. He was duly impressed by icebergs from Barry which were 10 to 20 times the size of his boat. It remained for the Harriman Alaska Expedition in 1899 to find a route past Barry Glacier and discover Harriman Fiord.

Vegetative studies show that during its last advance Barry Glacier did not ride up on Pt. Doran or completely block Harriman Fiord. In fact, the closer Barry Glacier pushed its terminal moraine towards Pt. Doran, the more it decreased the channel through which the tide flowed, and the more it increased the speed of the current and melting power of the sea at the face of the glacier, thus inhibiting its own advance.

When Barry began to retreat, the retreat was rapid and drastic. From its extended position within 1000 ft. of Pt. Doran, Barry retreated over three miles in fifteen years. By

An Observer's Guide to

Fig-61. In 1898, Barry Glacier (lower) extended a short ways into Harriman Fiord. It did not ride up on Pt. Doran (left side). Cascade Glacier (middle) was a tributary. Photo from Harriman Expedition Collection, Alaska and Polar Regions Dept. U. of A.

Fig-62. Barry retreated to near its present position by 1914. Photo by N. Simmerman '85.

Dora Keen's visit in 1914, Barry, Cascade, and Coxe Glaciers occupied their present approximate positions.

One can still see the effects of Barry's last advance on the vegetation along either side of Barry Arm. The Harriman Expedition noted that Barry had begun to retreat as there were barren zones on both sides of Barry Arm above which stood remnant forests. They observed many trees killed by the glacier's chill at the edge of the forest but few overturned trees. The trimline between the old and new forest is still visible (Fig-32).

On the west side of Barry Arm across from Point Doran, there is a flat, tidal marsh area with numerous dead trees. This area, which is the former outwash plain for Barry Glacier, has numerous kettles and push moraines. Saltwater intrusion killed the trees and started a new ecosystem — the tidal marsh — following the 1964 Earthquake when Barry Arm and Harriman Fiord subsided 6 to 8.1 feet.

Since 1914, the glaciers have been mostly in a retracted stable position building their moraines before undergoing another advance down the arm. Visitors have noticed small fluctuations in the Barry tidal ice-cliff which periodically covers and uncovers a rock ledge. In the seventies, the rock ledge was covered. In August of 1984, we noted a small dark area near tideline that looked like a rock ledge beginning to emerge. Following the annual winter advance, it was not visible in June of 1985. By late July, however, two

Fig-63. Since 1914, Barry Glacier has periodically advanced and retreated over these rock ledges. Harbor seals and sea otters often haul-out on the icebergs. Photo by Nancy Simmerman, August 1985.

dark areas were present. By mid-August of 1985 an unmistakable rock ledge was visible in several places. In June of 1986, we found that the annual winter advance had failed to recover the ledge and that more of it was now exposed. Whether the ledge will eventually emerge as a group of rocks, an island, or a point or will again be covered by a minor advance is one of those intriguing questions that keeps glacier-watchers returning year after year.

Changes are also evident at the medial moraine between Cascade and Barry which terminates to the west of the rock ledges. Between 1966 and 1969, it shifted position dramatically, as Cascade Glacier began to advance covering a rock cliff exposed along the southern side of its terminus. A new terminal moraine lies just beneath the surface to the west of the medial moraine. It is clearly visible at low tide just to the west of the medial moraine near the subglacial stream.

Black-legged kittiwakes nest on the cliffs near both Cascade and Coxe glaciers. In some cases, they actually build their nests on the ice.

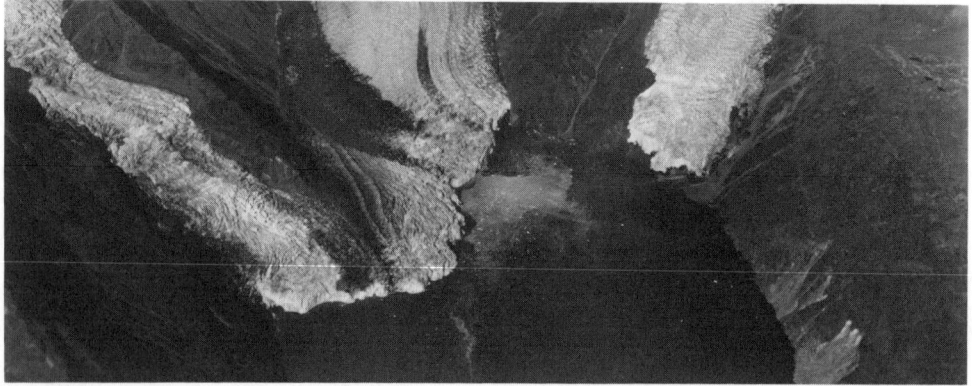

Fig-64 (top): Left to right: Cascade, Barry, Coxe. Note rock cliffs at Cascade's terminus and large rock avalanche debris from '64 Earthquake on Barry. Fig-65 (below) shows Cascade has advanced over rock cliffs. Post 1966 (photo damaged), 1969. USGS.

The Glaciers of Prince William Sound, Alaska

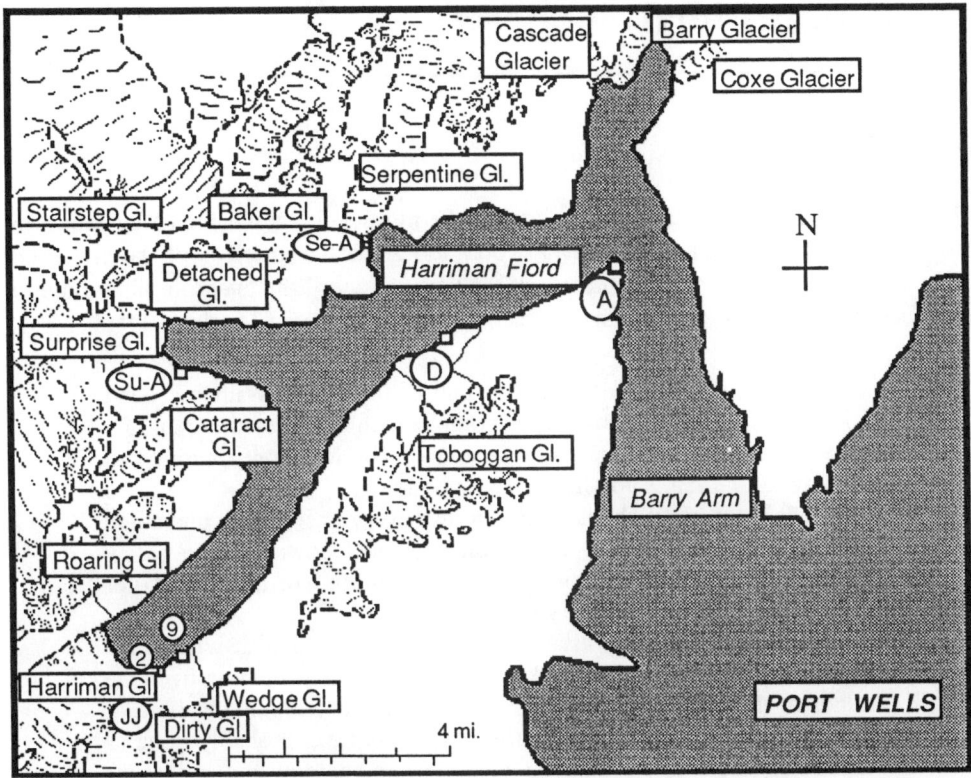

Fig-66: Barry Arm and Harriman Fiord. Sources: Field, *Survey and Photographic Stations in Harriman Fiord and Barry Arm, Prince William Sound, Alaska, 1964.* USGS, Anchorage A-4. 1960 and Seward D-4 Alaska, 1952, Cf. also Fig-41 for radiocarbon dates and sites.

Photographic Stations for Barry Arm and Harriman Fiord:

Station A, atop the cliff at Pt. Doran about 65 to 80 ft. above tideline, affords good views of Cascade and Barry Glaciers. A small cairn marks the spot. The station was first occupied by Grant and Higgins in 1905, then reoccupied by them in 1909, by Tarr and Martin (1910), Dora Keen (1925), and Field (1931, 1935, 1961, 1964, 1974, and 1976.) The best close-up views of the position of the terminus can be taken from a boat.

Station D is directly across Harriman Fiord from Serpentine Glacier. It is on an exposed ledge with sparse vegetative cover that juts out from the forested shoreline at an altitude of 15 to 30 ft. E.S. Curtis, photographer on the Harriman Expedition (and later of considerable fame for his photographs of American Indians), photographed Serpentine Glacier (Fig-68) from here. Subsequently, virtually every scientific party has reoccupied the station. There are also good views of Surprise, Detached, Baker and Penniman Glaciers.

Su-A was used by Field on his trips between 1957 and 1964. It affords good views of the terminus of Surprise as it advances. Station 9 is on the point just to the south of the mouth of the stream from Wedge Glacier. Station 2 is on the next point towards Harriman. Station JJ is on the prominent knoll on Harriman's east side, cf. Fig- 74.

An Observer's Guide to

Harriman Fiord

Discovery:

The discovery of Harriman Fiord ranks among the most exciting and foolhardy events in the exploration of Alaska. Fortunately, two of the greatest naturalists of the era, John Muir and John Burroughs, were present at the time to record the event. In his tribute to Harriman, John Muir describes Barry Glacier and the discovery of Harriman Fiord.

> As we approached the head of one of the Prince William Sound fiords [Barry Arm] it seemed to be completely blocked by the front of a large glacier and an outreaching headland. The local pilot, turning to our Captain Doran, said: "Here, take your ship. I am not going to be responsible for her if she is to be run into every unsounded, uncharted channel and frog marsh." The Captain slowed down, and in a few minutes stopped, after creeping forward to within half a mile or so of the front of the ice wall [Barry Glacier].
>
> Then Mr. Harriman asked me if I was satisfied with what I had seen and was ready to turn back, to which I replied: "Judging from the trends of this fiord and glacier, there must be a corresponding fiord or glacier to the southward, and although the ship has probably gone as far as it is safe to go, I wish you would have a boat lowered and let me take a look around that headland into the hidden half of the landscape."
>
> "We can perhaps run the ship there," he said, and immediately ordered the captain to "go ahead and try to pass between the ice wall and headland [Pt. Doran]." The passage was dangerously narrow and threatening, but gradually opened into a magnificent icy fiord about twelve miles long, stretching away to the southward. The water continuing deep, as the soundline showed, Mr. Harriman quietly ordered the captain to go right ahead up the middle of the new fiord. "Full speed, sir?" inquired the captain. "Yes, full speed ahead." The sail up this majestic fiord in the evening sunshine, picturesquely varied glaciers coming successively to view, sweeping from high snowy foundations and discharging their thundering wave-raising icebergs, was, I think, the most exciting experience of the whole trip. (John Muir 1911, pp 4-5.)

The return trip was even more exciting. John Burroughs records a few tense moments in his "Narrative of the Expedition." "On coming out of the inlet and turning almost at right angles into Port Wells, the tide which was with us and which was running very strong, caught our vessel and for a moment held her in its grasp. She hesitated to

Fig-67. After rounding Pt. Doran and slipping past Barry Glacier, the Harriman Expedition discovered Harriman Fiord. Note that Cataract Glacier (left) is still tidal. Surprise is much closer to Cataract than it has been seen since. Baker, a hanging glacier, extends much farther down the mountainside than in recent times. And, Serpentine reaches almost to Harriman Fiord. From the Harriman Expedition Collection, Alaska and Polar Regions Dept. University of Alaska, Fairbanks.

respond to her helm, and was making direct for the face of the great glacier on our port side; but presently she came about, as if aware of her danger, and went on her way in less agitated waters." (Burroughs 1902, p. 73.)

The vast amount of calving activity reported earlier by Glenn probably signalled the beginning of Barry's retreat, since tidewater glaciers retreat rapidly once they abandon their moraines through increased calving. The Harriman Alaska Expedition arrived at a slack time in Barry's activity and made their dangerous, foolish and subsequently famous passage through the narrow gap. They could have easily grounded on the moraine. Had they grounded, their ship, the *George W. Elder,* could have been exposed to large waves generated by the calving that were capable of hurtling icebergs against the ship's unprotected hull. Obviously, if this had happened, rescue services would not have been available. As it was, they escaped with only a bent prop, necessitating a return to Orca (Cordova area) for repairs; during the time taken for repairs, a scientific party including John Muir, Henry Gannet, and possibly E.S. Curtis had the opportunity to remain ashore in Harriman Fiord for several days.

The trip up Harriman Fiord itself was a highlight of the cruise. John Burroughs in his "Narrative of the Expedition," records the event and the naming of the glaciers:

> We went on under a good head of steam down this new inlet where no ship had ever before passed. It was one of the most exciting moments of our voyage. We could see another huge glacier about ten miles ahead of us with its front on the water barring the way. Glaciers hung on the

steep mountain sides all about us. . . . The scene was wild and rugged in the extreme. One of the glaciers was self-named Serpentine by reason of its winding course down from its hidden sources in the mountains — a great white serpent with its jaws glittering fangs at the sea. Another was self-named the Stairway, as it came down in regular terraces or benches. A Colossus of Rhodes with seven-league boots would have been an appropriate figure upon it [renamed Surprise Glacier]. As we neared the front of this last glacier the mountains to the left again parted and opened up another new arm of the sea, with more glaciers tumbling in mute sublimity from the heights, or rearing colossal palisades across our front. Another ten-mile course brought us to the head of this inlet, which was indeed the end of navigation in this direction. (Burroughs 1902, p. 72)

Harriman Fiord's Glaciers are discussed in the order in which they are seen as boats cruise up the fiord: Serpentine (west side), Toboggan (east side), Baker (west side), Surprise Inlet with Cataract (west), Detached (north) and Surprise Glaciers, and finally, upper Harriman Fiord with Roaring (west), Wedge (east), Dirty (east) and Harriman.

Fig-68. Serpentine Glacier from Station D. Note the prominent lateral moraine described by Tarr and Martin to the right of the glacier. Photo by E.S. Curtis (1899), Harriman Alaska Expedition Collection, Alaska and Polar Regions Dept. U.of A., Fairbanks.

Serpentine Glacier:

Location: Harriman Fiord
Distance from Whittier: 34 miles
Access: Boat
Type: Valley glacier descending from cirque basin and icefield

Area: 32 sq. miles (Post 1986, unpub. work)
AAR: 83% (Post 1986, unpub. work)
Aspect in accumulation area: SSW, at terminus: SSE
Status: Stable, building terminal moraine
Photos: Figs-67-69
Maps: Figs-41, 66

Most of Serpentine Glacier has its accumulation area in Mt. Gilbert's snowfields and cirques but a small portion originates from an icefield. The same icefield feeds Colony Glacier, which flows into Lake George on the western side of the Chugach Mountains in the vicinity of the Matanuska Valley. A tributary glacier flowing from Mt. Muir enters Serpentine about 1/2 mile from the western terminus. From the cove, one can see that at least one of Mt. Gilbert's tributaries no longer reaches Serpentine Glacier, while the very white tributary just to the SE now flows and melts on top of Serpentine

Fig-69. Serpentine Glacier, September 23, 1981. Small islet marks the position of the terminal moraine in 1870's. Photo by Austin Post, USGS.

without reaching its terminus. Much debris from medial and ablation moraines covers Serpentine's surface; this is more apparent from the air than from the cove. The two apron glaciers clinging to the side of Mt. Muir above Serpentine Cove are East and West Penniman Glaciers. They are slowly shrinking and often have ice avalanches.

The cove itself hangs hundreds of feet above Harriman Fiord's much deeper trough and is separated from it by its 1870s, lobate moraine. When Serpentine extended to this moraine, its ice-cliff was 2 1/2 miles wide.

Serpentine's tidal eastern side has made several short advances and retreats during the last 85 years, but it has never regained the advanced position it occupied on the terminal moraine abutting Harriman Fiord in the 1870s. The rock ledge near the eastern terminus was barely visible in 1910, covered sometime before 1938 and uncovered between 1961 and the early seventies. By 1984, there were small alders, lichens, annual plants on the ledge. Striation marks are still clearly visible on the rocks.

Serpentine's western margin formerly extended much farther south and was covered with a very heavy coat of ablation moraine. Tarr and Martin (1914) found the small flat in front of the glacier to be underlain with blocks of black ice — the first step towards the formation of kettles. They attributed the western side's more advanced terminus to the protection afforded by the ablation moraine and to shading by the mountain. Austin Post's aerial photo shows that the western margin is still more advanced than the tidal-washed eastern face. There is a brief barren area in front of it; a forest now covers all of the old 1870s outwash plain. Several small kettle-type lakes lie hidden in the forest.

The old eastern lateral moraine, dating from when Serpentine was a much larger glacier filling the cove and extending 2 1/2 miles along the edge of Harriman Fiord, is clearly visible on the hillside above the cove. Tarr and Martin described the lateral moraine on the eastern side as: ". . . one of the best developed deposits of this sort which we have seen in Alaska. It is a narrow-crested ridge 30 to 50 feet high and with so sharp a crest that it is difficult to walkupon it. On the western side of this ridge is a bare slope of ground moraine and bed rock extending to the shores of the cove, and to the present border of the glacier; on the eastern side it is bordered by dense mature forest. This lateral moraine is a conspicuous object in the region, curving northward along the margin of Serpentine Glacier for a distance of over a mile, beyond which it could not be seen from sea level (Tarr and Martin 1914, p. 329)." In 1985, we found this moraine to still be a dominant feature of the sky line, although it now supports a thick covering of alders and conifers. The area between the moraine, cove and glacier is densely vegetated. Another prominent lateral moraine runs along the western side above the glacier.

In 1964, the land subsided between 7.2 and 7.9 feet. Along the shoreline are many trees killed by saltwater intrusion. Field predicted that the increased exposure of the tidal ice-face to saltwater would augment the rate of ablation at the terminus causing the glacier to retreat. At present, the face is protected by a 5 ft. high push moraine atop its terminal moraine. It appears to be stable.

Fig-70. Toboggan Glacier. Photo by Sidney Paige, USGS. Aug. 20, 1905. Photo damaged.

Toboggan Glacier (eastern side):

Toboggan Glacier flows from a low-lying icefield that caps the peninsula between Port Wells and Harriman Fiord and squeezes down a narrow canyon opposite Serpentine. Nearly all of its accumulation area lies below the 3000 ft. level. As one would expect, it has retreated in response to the general warming trend following the end of the last neoglacial period. Grant and Higgins using vegetative studies found that Toboggan Glacier had been tidal during the late 18th century. When first seen by the Harriman Expedition, it was no longer tidal. During this century, it has continued to shrink in volume and retreat. Toboggan's main tributary lost ground from 1935-1957 and separated from the glacier in 1964. In his preliminary study (1980d) of Harriman Fiord, Post shows its terminus at the 1000 foot level. Observations made from a boat suggest it is still retreating.

Baker Glacier (western side):

Named for Dr. Marcus Baker (1849-1903), editor of the *Geographic Dictionary of Alaska*, Baker is fed by cirques on Mt. Muir (5000 to 6000 ft. elevation), which in turn accumulate snow from the precipitous avalanche slopes that tower 1315 to 2630 ft. above them. Baker Glacier terminates about a 1000 ft. above tidewater.

The Harriman Alaska Expedition reported that Baker reached the base of the cliff above the outwash plain. By 1905, it had retreated up the cliff; shortly after 1905, Baker began to advance rapidly. In 1914, Dora Keen reported it 1000 ft beyond its 1910

Fig-71. When first photographed by the Harriman Expedition, Baker Glacier extended to the bottom of the rock cliffs in the left hand corner. Photo by Nancy Simmerman, 1985.

position. Sometime between 1914 and and 1931, Baker retreated to behind its 1905 position (Field 1975), then made a slight advance between 1931 and 1935. Since 1935, Baker has changed very little.

Baker's most striking features include a series of rock piles resembling medial moraines, that emerge mysteriously from the middle of the glacier. They have probably been formed by the glacier plucking away at an ice-concealed, bedrock knoll.

Surprise Inlet:

Surprise Inlet was just barely opening up when the Harriman Alaska Expedition visited. At that time, Surprise and Cataract Glaciers were joined. Detached Glacier was named by Grant and Higgins, who found that Surprise Glacier had retreated seven-tenths of a mile in six years and become separated from Cataract and Detached Glacier.

How far did Surprise Glacier advance out into Harriman Fiord? Depths shown on Post's preliminary bathymetry of Harriman Fiord and Surprise Inlet suggest that a sill or old moraine extends from the Bull Pens in an arc to the point between Surprise Inlet and Harriman Fiord. Tree-ring studies place its last glacial maximum at about 200 years ago (Post, personal communication.)

The Glaciers of Prince William Sound, Alaska

Fig-72. Between 1899 when the Harriman Alaska Expedition found Cataract (left) and Surprise glaciers just barely touching and Grant and Higgins' return in 1909, Surprise Glacier retreated to near its present position, opening up a new bay appropriately called "Surprise Inlet." Surprise Glacier remained nearly stable until the early 1980s when it again began to advance. Photo by Nancy Simmerman, 1985.

Cataract and Detached Glaciers:

As one approaches Surprise Inlet, Cataract Glacier appears to rest in a hanging valley about 2000 ft. above the Inlet on the eastern side. However, its accumulation area, although small, extends to the 5000 foot level where it is fed by snow avalanching and blowing from the surrounding peaks and ridges.

When first seen by the Harriman Expedition, Cataract Glacier reached tidewater barely joining Surprise Glacier. It ceased to be tidal sometime between 1925 and 1931, then regained its tidal position in the advance of 1935 (Field 1935a, 1937). Cataract receded again in the late '30s and has not been tidal since. Field estimated that in 1968 it was about 250 to 330 feet above sea level (Field 1975). In 1985, we estimated the lower tongue at about 800 feet and the upper one at 1500 ft. or more. Numerous ice avalanches broke from its face temporarily whitening the cliffs below.

Formerly, Detached Glacier (west side) descended from cirques on Mt. Muir (named for John Muir, founder of the Sierra Club) to join Surprise Glacier. However, Grant and Higgins (1909) found it already separated from Surprise and terminating about 1/4 mile from the shore. By 1977, its terminus had retreated from 133 ft. (1909) to approximately 2100 ft. above sea level. Several waterfalls plummet from the glacier to the fiord through a deep-cut gorge. Ice avalanches are sometimes frequent (Fig-17.)

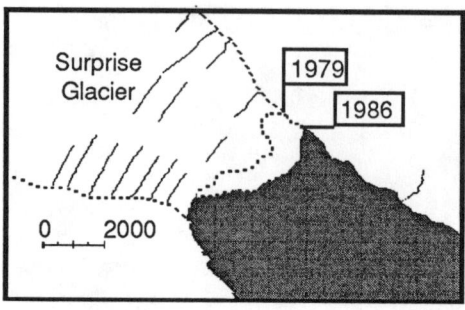

Fig-73: Terminus Positions of Surprise Glacier: 1979 - 1986.

Sources:

Post, A. 1980d. Terminus in 1979.
Post, A. Aerial photograph, USGS. Sept. 1986.

Surprise Glacier:

Location: Head of Surprise Inlet
Distance from Whittier: 38 miles
Access: Boat
Type: Ice-calving valley glacier descending from icefield
Length: 8 mi. (Field 1975)
Area: 32 sq. miles (Field 1975)

AAR: 83% (Post 1986 unpublished work)
Aspect: ENE glacier as a whole; ESE at terminus
Status: Between 1979 and 1986, the east side remained stable while the center and west side advanced approximately 800 ft.
Photos: Figs-68, 72.
Maps: Figs-41, 66, 73.

Surprise Glacier is the most beautiful glacier in Harriman Fiord as it winds 8 miles down a steep-walled canyon to terminate in domineering ice-cliffs at the head of Surprise Inlet. Its prominent medial moraine is visible from Pt. Doran. The sources for Surprise Glacier lie in an accumulation basin that also feed Colony and Lake George glaciers.

Like other glaciers in Harriman Fiord, Surprise underwent a retreat coinciding with the end of the last neoglacial period. Between 1898 and 1909 it retreated rapidly from a position near Cataract Glacier to its present position (1.24 miles) (Grant and Higgins 1913). For most of this century Surprise Glacier has been in a retracted stable position building a new moraine at the head of its fiord. The most significant feature for gauging its position has been a rock ledge, first reported by Grant and Higgins, that projected from the front of the glacier near its south side. According to Field, this divided rock ledge has been visible in all photos taken since 1909, indicating the glacier's recent stability (Field 1975). The ledge was sufficiently exposed to provide an ice-free rookery area for a colony of black-legged kittiwakes, which numbered 514 nests in 1972. In 1966, Surprise Glacier's northern tributary ceased to flow into Surprise and became an independent hanging glacier (Field 1975).

However, recently Surprise Glacier has begun to advance. In 1985, we found the rock ledge subject to ice-avalanches, and the kittiwake rookery abandoned. Post's 1986

Fig-74. Harriman Glacier has been advancing for several centuries. Its terminus is now approaching the stream from Dirty Glacier. Photo by Nancy Simmerman.

aerial photos taken in September suggest that the glacier had advanced over the rock ledge. In August of 1987, we found the rock ledge completely covered and evidence that the glacier was now also slowly pushing up a carpet of tundra along its eastern edge.

Return to Harriman Fiord:

The upper end of Harriman Fiord has been ice-free for the past 2300 years (i.e. since 300 B.C.). Radiocarbon dates show that the area just south of the point separating Harriman Fiord and Surprise Inlet was ice-free about 200 years after Barry Glacier retreated from the moraine between Hobo Bay and Pakenham Point (Heusser 1983).

Except for Harriman Glacier, which has been advancing for several centuries, all the other glaciers here have responded to the end of the neoglacial period by retreating.

Roaring, Wedge, and Dirty Glaciers:

Roaring Glacier, named by the Harriman Alaska Expedition for the roaring sound from its numerous ice avalanches, is now a small hanging glacier located 3/4 mile north of Harriman Glacier on the northwestern side of the fiord. During the first half of the century, Roaring retreated slowly and shrank; since mid-century it has been stable.

Wedge Glacier, whose name draws attention to its wedge-shape which contrasts with the typical bulb-shaped termini, occupies the crest between Bettles Bay and Harriman Fiord. Its icefield at 2700 ft. is well below the snowline of most Alaskan coastal areas. Since the Guggenheim Alaska Expedition of 1909 visited Harriman Fiord and went ashore to visit Wedge Glacier, the glacier has retreated significantly. Today, its narrow wedge-shaped terminus lies at about the 800 foot level.

Dirty Glacier (3/4 mile east of Harriman's terminus) hovers beneath a thick, black ablation moraine. Since first observed, Dirty Glacier has retreated and shrunk. In 1910, it was close to tideline; by 1987 its terminus was well back from the shore. Many small kettles dot the outwash plain.

Harriman Glacier:

Location: Head of Harriman Fiord
Access: Boat
Type: Ice-calving valley glacier descending from an icefield
Length: 8 miles (Field 1975)

Area: 19.7 sq. miles (Field 1975)
AAR: 68% (Post unpublished data)
Slope/aspect: Northeast
Status: 1981-1986: advanced along sides about 50 ft.; filled in two large embayments representing over 500 ft. advance in these areas.
Photos: Figs-74-75. Maps: Figs-41, 66.

Harriman Glacier lies at the head of Harriman Fiord; its broad terminus, numerous coalescing tributaries and long, low-grade ascent for 10 miles towards the distant peaks make it one of the most impressive glacial vistas in the sound. It is also interesting from a glaciologist's point of veiw, because it has one of the lowest altitudes for an accumulation area in the region. The firn is estimated to average between 1600-1700 ft.(Post, personal communication), which is about the lowest of any place along the southern and southcentral Alaskan coastline. The icefield is so thin that one can see the contours of the underlying terrain, yet only a few peaks penetrate the snow-cover. Harriman's icefields also drain into Surprise Glacier (north), Twenty-mile glacier (west), Billings, Pigot and Bettles glaciers (south), and Dirty Glacier (east).

How long has it been since Harriman Glacier retreated from the land that it is now reclaiming? The various answers given to this question throughout the century illustrate the advances in technology that glaciologists have been able to apply to their studies. When the Harriman Expedition (1899) first observed the glacier and its environs, they estimated that it had not been farther advanced since the 16th century. This figure was refined by Viereck's tree-ring analysis (1936), which established that it had not been more

Fig-75. E.S. Curtis made the first photograph of Harriman Glacier (southern side) in 1899. Photographic Station JJ is on the 748 ft. knoll to the left.

advanced since at least 1525 A.D. Subsequent radiocarbon dates (Heusser, 1983) indicate that areas on the knoll adjacent to its eastern terminus (near station JJ) have not been covered for 1900 years (Fig-41).

Throughout this century, Harriman Glacier has been advancing — slowly but steadily. By 1971, Harriman's south side had advanced 3950-4275 ft. beyond its 1899 position, and the northern side was 2460-2800 ft. more advanced than in 1899. In 1979 and 1981 two large embayments, which in places were more than 500 feet back from the terminus, developed, possibly threatening Harriman's advance. However, between 1981 and 1986 Harriman refilled both embayments and advanced about 50 feet along its margins. Thus its southern side has advanced almost 7/8ths of a mile in the past 88 years, while the northern side has only advanced approximately 1/2 mile. Because of the advance, a small kittiwake rookery near the glacier was abandoned in the early 1980s.

History:

During the height of the gold prospecting era, Harriman gained some notoriety. Its slow but persistent advance overrode a gold claim between the time prospectors decided to develop it and the time their supplies arrived a few months later!

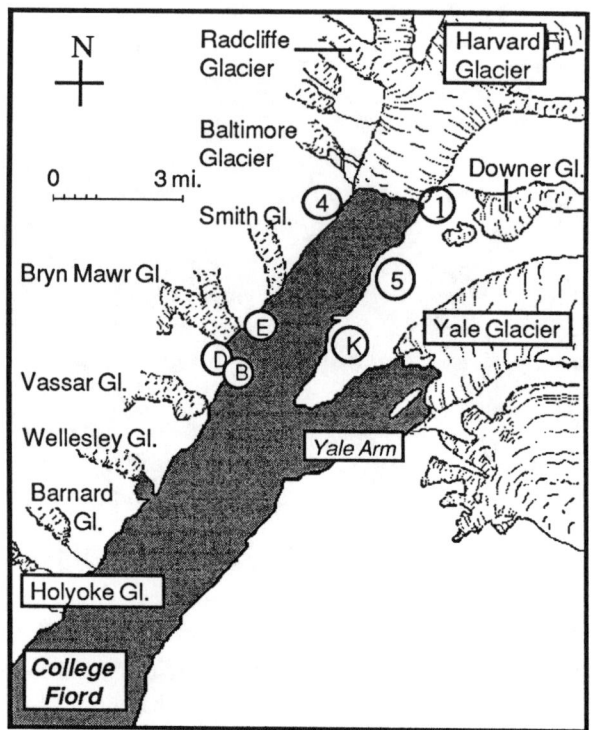

Fig-76. Glaciers of College Fiord and Yale Arm.

Sources: Field, *Survey and Photographic Stations in Upper College Fiord. Prince William Sound, Alaska 1964. A.G.S. Glacier Surveys.* Map No. 64-3-G10.

USGS Anchorage, A-2, A-3, B-2 and B-3, Alaska 1960, 1:63,360 (based on aerial photography taken 1950 to 1957).

Photographic Stations:

Large discharges of ice can play havoc with plans to occupy photographic stations near Harvard Glacier. The best photographic point is Station 5 atop Field Island (named by Post), an islet a short distance from the eastern shore about a mile from the ice-front. One can also photograph Baltimore Glacier, a small glacier to the west of Harvard. Sta. 5 is near stations occupied by Grant and Higgins in 1909, Tarr and Martin in 1910, and was used by Field in 1935, 1957, 1961, 1964, 1966, and 1976. A cove lies to the south of Field Island. Austin Post (personal communication) reports exploring this with the research vessel *Growler*. He found the cove too shallow for anchoring, but recommends anchoring off its entrance for short stops but cautions that a watch should remain aboard in case ice moves down on the boat. The ice hazard is too great for long stays. The next best station is Field's Station 4 on the west side of Harvard Arm. According to Field (written communication, April 16, 1986), the station is on an obvious good viewpoint about 400 to 500 ft. above a cove where one can land. A cairn marks the spot. In 1976, the cairn was at the edge of the alders. If there is too much ice to reach Stations 4 or 5, Station K is the best substitute. Nearly 3.1 mi. from Harvard, Station K is located on the east side of Harvard Arm opposite Smith Glacier on a bold rocky headland with a small cove immediately to the south. The station is on a rock jutting out from the brush and is about 30 to 50 ft. above tideline. The Harriman Expedition probably used this station in 1899. Station K is also a good place from which to photograph Smith and Bryn Mawr glaciers. Stas. A to E are by Bryn Mawr. A and C are on the southern hillside, D and B on points, and E on an islet.

Chapter 11. Of Advance and Retreat: College Fiord

College Fiord runs on a northeast/southwest axis in the northwestern part of Prince William Sound. At 61°15' N it penetrates farther north into the Chugach Mountains than any other fiord. All the glaciers in College Fiord, drain from the series of small interconnected icefields spanning the crest of the Chugach Mountains known collectively as the Chugach Icefield. This icefield feeds Meares, Columbia, Shoup, and Valdez. It also feeds the Matanuska, Nelchina, and Tazlina Glaciers to the north; Knik and Marcus Baker to the west, and Stephens, Klutina, and Tonsina to the east. A barrier sill/moraine separates College Fiord from adjacent Port Wells.

Because there are no ice-free anchorages beyond Coghill, College Fiord is seldom visited for any length of time by pleasure boats. However, the area is a popular destination for kayakers out of Whittier and cruise ships routinely visit. For boaters willing to tolerate a little ice in their anchorage, Post recommends the bay west of the overridden island in Yale Arm. A moraine blocks passage along the southern shore of the island.

As in the vicinity of all ice-calving glaciers, currents may pose a hazard. Grant and Higgins were the first to report currents near tidewater glaciers that flow towards rather than away from the glacier. At Harvard, they observed:

> Waves generated by the fall of icebergs and the strong currents in front of the glacier made it impractical to approach near the glacial front in a small boat. Reports are current that native seal hunters in bidarkas have been drawn under the glacier by northward flowing currents. At the time of our visit in 1909, there were marked northward-flowing currents on both sides of the fiord near its upper end. (Grant and Higgins 1913, p. 30)

Tarr and Martin (1910) added that waves from large calving events at the head of the fiord were clearly scouring out cliffs and uprooting shrubs along both sides of the fiord.

Topography and Climate:

Together, topography and climate play the major roles in the continuing drama of glaciers in College Fiord. The mountains on the southeastern side are lower than those on the northwestern side of the fiord. On the peninsula separating Unakwik Inlet and College Fiord, the major ridge-line lies between 2500 and 3000 ft. and only a few peaks are over 3500 ft. As a result, except for Yale Glacier, glaciers on the southeastern side are generally smaller than those on the northwestern side. The glaciers on the eastern side are retreating probably as a result of the climatic warming following the end of the last neoglacial phase. It seems likely that prior to this recent warming trend, the firn-limit lay well below the 4000 foot level giving these glaciers large, low elevation, accumulation areas. With the warming trend, the firn-limit rose, significantly reducing their accumulation basins and initiating retreats.

The situation is quite different on the western side. Here, a stunning series of spectacular glaciers tumble down sheer slopes from accumulation areas well above the critical 4000 ft. level. Many terminate near or at sea level. Their healthy, higher elevation accumulation zones receive ample snowfall; and, because of their rapid descent, very little of the glaciers' total area lies below the 4000 ft. level as compared to Yale.

Glacier	Rate of descent
Bryn Mawr (west)	3700 ft. to the mile
Vassar (west)	2500 ft. to the mile
Smith (west)	2200 ft. to the mile
Yale (east)	600-700 ft. to the mile

The 4000 ft. level appears to be the dividing line between areas which receive heavy rainfall during the sound's fall equinox storms and the higher slopes that receive snow. Warm rain melts glacial ice faster than sunlight. In College Fiord, whether a glacier receives warm rain or snow-falls during these autumnal gales may be the most critical factor in determining whether a land-terminating glacier is stable or retreating. For ice-calving tidewater glaciers, whether the terminus is in an extended or retracted position remains the important variable.

Slope aspect also plays a role on the opposite sides of the fiord. Glaciers on the eastern side receive the afternoon sun when its warmth and melting effects are greatest. By contrast, the sun strikes the glacial slopes on the western side in the morning when its rays are weakest. Thus, less melting occurs.

Because of this fortuitous combination of topography and climate, the series of glaciers along the western side is unique. Unlike nearly all other glaciers in the Prince William Sound region, they have shown relatively little drastic variation throughout this century. What fluctuations have occurred in the position of the termini are excellent illustrations of the cyclical nature of ice-calving glaciers.

Exploration:

College Fiord is first mentioned in Vancouver's journals. After naming Point Pakenham, Whidbey proceeded up College Fiord. He noted several concealed rocks, undoubtedly the moraine (cf. Fig-41), and considerable ice. The ice forced him to discontinue his exploration, but not before noting that the bay ended in more ice which extended into the mountains. Whidbey did not observe Yale Arm or Yale Glacier. (Vancouver 1798, 183.)

In 1898 as part of the U.S. Army search for a route to the interior, Capt. Glenn led a survey team on a reconnaissance mission of the sound's northern fiords. Although he failed to find an ice-free route, he left this description of Yale and Harvard Glaciers.

> The day was dry and clear. Directly in our front was the most imposing sight we had yet seen — I might add more imposing than any we saw during the season. Glistening in the sun were two large glaciers, which we named the "Twin Glaciers," [now called Harvard and Yale] the pair being separated by a short ridge or hogback that runs down to salt water. In front of the one on our right [Yale] the sea ice extended for over 3 miles, while in front of the other [Harvard] this sea ice [ice pans formed in the winter] extended at least twice that distance. This ice was covered by snow several feet in depth. . . we could make no headway against it with the boat. (Glenn 1899, p. 19)

The Harriman Alaska Expedition (1899) was the first scientific exploration of the fiord. They photographed, described and named glaciers on the western side for Womens' Ivy League Colleges and those on the eastern side for mens'.

The glaciers and their recent history are described in the order in which they are seen as one cruises up the fiord: Crescent and Amherst (east side), Barnard and Holyoke (west), Wellesley (west), Vassar (west), Yale (east), Bryn Mawr (west), Smith (west), Baltimore (west), Radcliffe/Harvard (head).

Coghill area, Crescent and Amherst Glaciers:

Several glaciers lie in the mountains to the east of the Coghill area at the mouth of College Fiord. Of these, Crescent and Amherst are the most readily seen by boaters. Neither glacier has been tidal for a very long time — probably not since the end of the Pleistocene. Their well-vegetated outwash plains are anomalous features for Prince William Sound where outwash plains are predominantly raw and unstable.

Crescent is in slow retreat as indicated by the newly exposed rock. Along the outer wall of the glacier's moon-shaped curve, snow and rock avalanches descend from the mountainside, weakened by persistent undercutting. Periodically, large rock avalanches have run out on to the glacier. One can trace the forward movement of the glacier by noting how far these avalanche tongues have been carried from their place of deposition.

Ice-calving Amherst Glacier descends 5 miles from large cirques near Unakwik Peak and is the largest glacier in the Coghill River watershed. It terminates in a picturesque lake, which makes it an attractive destination for hikers. From a boat, one can see prominent bergschrunds and the bumpy upper surface — a sign that the glacier has thinned enough to reflect the contours of the underlying bedrock. Dating based on vegetative studies (Viereck) show the moraine 1/2 mile in front of the glacier was its maximum attained about 1832. Reports by Tarr and Martin (1910) and Field (1975) indicate that Amherst has been retreating throughout this century. In 1985 barren areas around the glacier indicated continued retreat.

Holyoke and Barnard:

The first large glacier on the western side of College Fiord is Holyoke. South of it are several small apron glaciers. Like land-terminating glaciers in other parts of the sound, these have retreated off their neoglacial moraines. The retreat has been small, because the steep slope enabled the glaciers to rise quickly to a new elevation where they re-established their equilibrium.

Although apron glaciers are small and insignificant, they exhibit a few features not observed elsewhere. The glaciers are very thin and moving very slowly down the mountain. Because there is little to disturb the ice, it is possible with binoculars to see annual accumulation layers not only in the snow, but also in the ice. In August of 1985, the firn limit extended to the termini of these glaciers.

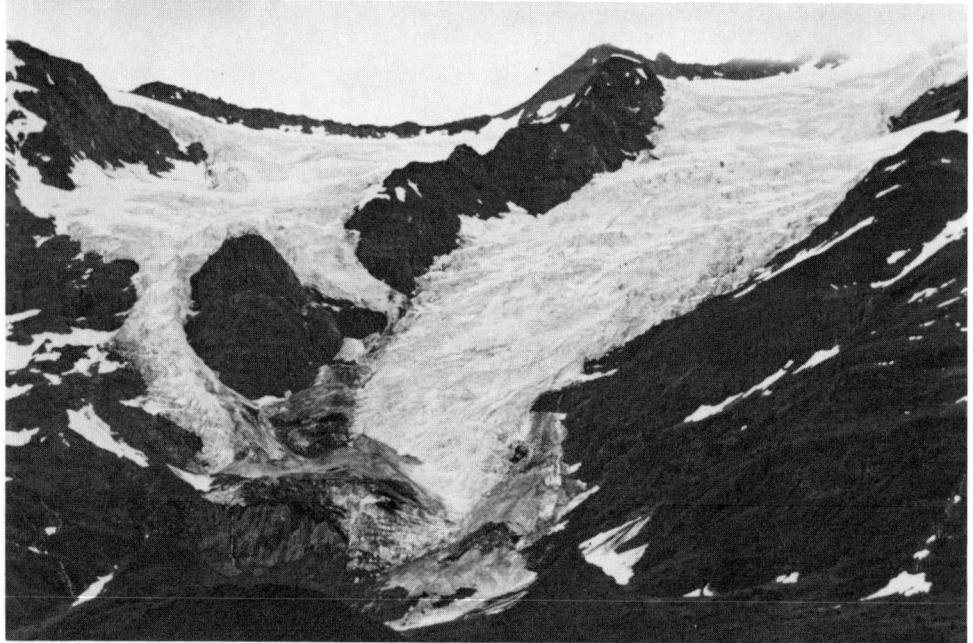

Fig-77. Holyoke is fed by two cirque basins. A neoglacial moraine is in the foreground. Photo by Nancy Simmerman. 1985.

Holyoke Glacier is fed by two cirque basins between which towers a prominent arête. Vegetation grows virtually to the edge of Holyoke suggesting that it has been in equilibrium for some time. Holyoke's most interesting feature is its old terminal moraine which cuts across the early neoglacial lateral moraines that lay along the slope of College Fiord. This suggests Holyoke advanced since the end of the recent neoglacial maximum.

The outstanding features of Barnard Glacier include its stairstepping pattern, extensive rock debris on the glacier, and longitudinal crevasses. Evidence for the developing of glacier stairstepping first occurs where the cirques at the base of the avalanche accumulation zone pour over icefalls into a basin. Indentations in the glacier suggest the possibility of a second basin still concealed above the first. In 1985, we noted extensive rock debris that may have originated from a large rockslide at the cirque headwall. At the terminus, well-developed longitudinal crevasses occur where the glacier spreads out in a bulb-shaped snout. Vegetation around the terminus indicates that Barnard has remained stable during much of this century or it may even be slightly advancing.

Wellesley Glacier:

Location: West side of College Fiord
Access: Boat
Distance from Whittier: 41 miles

Type: Valley glacier descending from cirque
Slope/aspect: Eeasterly
Status: Advanced approx. 600 ft. between 1981-1986

Wellesley lies at tidewater at the head of a 1/2 mile wide inlet, whose outer edge is protected by a moraine possibly dating from the 17th century. Like other glaciers in College Fiord, Wellesley advanced during the first two decades of the 20th century. In 1975, Field considered it one of the most stable glaciers in College Fiord, not changing more than 300-700 ft. in 70 years. As a result of this lengthy stability, Wellesley has built up a new moraine in front of its terminus. With its terminus protected by

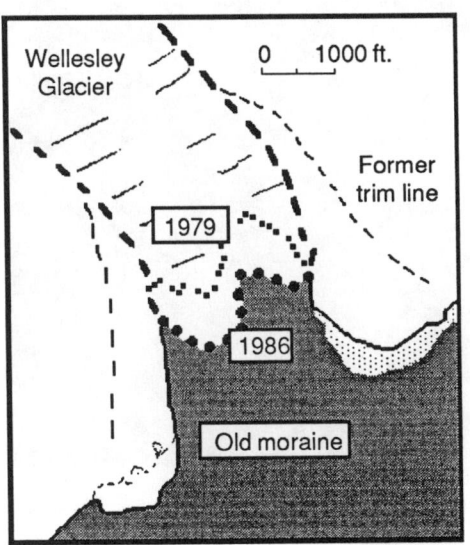

Fig-78. Terminus Positions of Wellesley Glacier: 1979-1986.

Sources:

A. Post,. 1980e. Terminus in 1979.
A. Post, . Aerial photograph, USGS, Sept. 1986.

the moraine from the salt water, Wellesley's melting has diminished relative to its accumulation, and the glacier has begun to advance, pushing its moraine across its shallow tidal basin. The old outwash plain has been stabilized by felt-leaf willows; in 1985, Sitka spruce were just beginning to overtop the willows. If the advance continues, these trees may be knocked over by the glacier before the end of this century.

Vassar Glacier:

Location: West side of College Fiord
Access: Boat
Distance from Whittier: 43 miles

Type: Valley glacier descending from a cirque
Length: 2 1/2 miles
Slope/aspect: Easterly
Status: Stable

A heavy ablation moraine, so thick that even the tidal ice-cliff is black, distinguishes Vassar from other glaciers in College Fiord. Tarr and Martin (1914) found that the black ice-tongue, usually indicative of a glacier in retreat, was advancing over willows and alders near sea level. What amazed them was that incredibly no glacier-ice showed through the heavy cover of rocks and debris. So thick and stable was the ablation moraine that they observed flowers and mosses growing on it.

Vassar Glacier has advanced twice during this century reaching maximums between 1914 and 1931 and around 1950 (Field 1975). Vassar advances until its terminus reaches the end of its shallow basin which hangs above College Fiord. Once the terminus pushes its protective moraine over the cliff, the glacier's melt-rate increases, and it retreats. Retreat is brief, since the precipitous mountain slope allows the glacier to quickly reach a new state of equilibrium.

Fig-79. The terminus of Vassar Glacier is covered by a thick ablation moraine. Photo by Nancy Simmerman 1985.

The Glaciers of Prince William Sound, Alaska

Fig-80. When Gilbert studied this photograph taken by Mendenhall in 1898, he spotted a rock ledge just beginning to emerge from under Yale Glacier (just to the right of center). Yale Glacier has now retreated about 3 miles from its 1898 position. Photo by Mendenhall. 1898. World Data Center Collection.

Yale Glacier:

Location: Head of Yale Arm, eastern side of College Fiord
Access: Boat and cross-country walk from tidewater, no trail.
Distance from Whittier: 47 miles
Type: Valley glacier descending from icefield
Length: 21.7 miles (Field 1975), now less.
Area: 74 sq. miles (Post 1986, unpub.)
AAR: 81% (Post 1986, unpub.)
Slope/aspect: Westerly
Status: Drastic retreat
Photos: Fig-80, 82
Maps: Figs-76, 81

Yale Glacier flows from the same icefield as Columbia Glacier. Many of its tributaries descend from the western side of Mt. Castner and other peaks between College Fiord and Unakwik Inlet. The Dora Keen Range (named by Post) with Mt. Glenn (9,806 ft.) and Mt. Witherspoon (12,012 ft.) separates it from Harvard Glacier. Yale is one-and-a-quarter to two miles wide, which makes it wider but not as long as Harvard. Its tidal ice-cliff actively calves filling the fiord with many small and moderate sized bergs. It is not always possible for small boats to enter the Arm.

Recent History:

In 1898, the U.S. Army was interested in finding a faster and safer route from the sound over the Chugach Mountains to the interior gold fields. Lt. Castner reportedly led a party on a climb 12 miles up Yale in hopes of finding an alternative to the Valdez Glacier route to the interior gold fields. Castner seems to have been more interested in the glacier itself than in any possible route over it. But, perhaps this is because Yale Glacier led only to an icefield, and not to open country amenable to roads and commerce.

> On a former trip I had climbed the right-hand glacier at the head of Port Wells, and from its summit looked far into this region. The glacier was apparently a relic of the glacial period, for, though of great extent, only a few streams flow out of it. Its ice and snow seemed perpetual. The climb to the summit of this glacier permitted us a view of its interior. As we broke trail on our snowshoes up the steep incline of snow and ice between the mountainside and the side of the glacier, we could see great masses of snow and ice thrown and piled together in every conceivable position. The lower masses of ice were almost as dark as the rock about them. They had been traveling neighbors for centuries, ground down and carried along by billions of tons of their younger fellows above. We narrowly escaped several snowslides while indulging our curiosity about the make-up of glaciers.
> (Castner 1900, pp. 686-687)

Mendenhall first photographed Yale in 1898. In June 1899, Gannett, Merriam, and Curtis photographed College and Harriman Fiord's glaciers for the Harriman Expedition. Gilbert, the expedition's glaciologist, studied these photographs and concluded that ". . . a blackening, west of the middle, by englacial drift suggests that a rock knob may lie near the surface, ready to develop into a nunatak or island if the glacier shall diminish (Gilbert 1910, p. 83)." The ledge, which became an island, has been uncovered since 1961. By 1986, Yale had retreated almost 3 miles from its 1898 position.

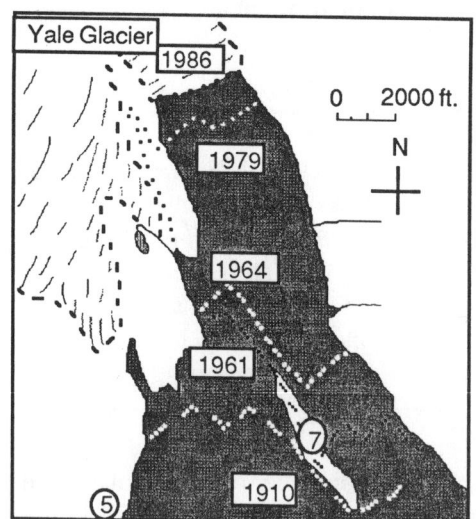

Fig-81. Positions of Yale Glacier's Terminus from 1910 to 1986.

Sources:

Field, *A.G.S. Glacier Studies, Map No. 64-3-G7*. Post, 1980e. for 1979 position. Post, aerial photograph Sept. 1986.

Photographic Stations:

The best historical viewpoint is Sta. 5 on the northwestern shore of Yale Arm. It is located on a ledge which juts slightly out into the inlet a few yards above the tideline and is marked with a cairn. All major parties have used it since 1899. The best modern station is No. 7 on the island. It was established by Field in 1964. A new station is needed nearer the glacier.

On the basis of earlier photographs and their own observations, Grant and Higgins (1913) surmised that Yale Glacier had reached its maximum within the last century, was now retreating, and had not extended to the mouth of the fiord in several centuries. Viereck's later tree studies established that Yale had covered the barren area between 1807 to 1827, but it had not extended further towards the mouth since at least 1650 (1967).

A very prominent lateral moraine, formed at the same time as Harvard Glacier's advance which ended about 3000 years ago, runs along the ridgetop on the southeastern side of Yale Arm. A good day-hike to the top of this ridge provides excellent viewing of Yale Glacier's surface features and College Fiord.

Evidence of former push moraines can be seen along the shores on both sides of the glacier. The recent retreat of Yale has been accompanied by shrinkage which has also left barren areas above the glacier.

Fig-82. Yale Glacier entered the drastic retreat phase of its cycle in the early 1960s. This photo by Post (USGS) in 1963 shows Yale's southern terminus still resting on the island.

Bryn Mawr Glacier:

Location: Western side of College Fiord
Access: Boat
Distance from Whittier: 45 miles

Type: Valley glacier descending from cirque
Length: 5 miles (Field 1975)
Slope/aspect: Southeasterly
Status: Approx. 200 ft. (or less) advance between 1979 and 1986

Bryn Mawr is formed by two tributaries which unite about a mile from tidewater. Like its sister glaciers, Bryn Mawr descends in giant stairsteps down the mountainside.

In 1909, Grant and Higgins found Bryn Mawr Glacier the "largest and most attractive glacier of those on the west side of College Fiord. It is a veritable ice cascade and gives a vivid impression of a rushing torrent to one who views the glacier from a point directly in front and not far distant." (Grant and Higgins 1913, p. 31)

When Tarr and Martin returned a year later, they found this most attractive glacier rampaging down the mountainside crushing shrubs and overriding spruces up to 5 inches in diameter. Avalanching ice-blocks destroyed vegetation 25 ft. away from the moraine. The advancing push moraine overrode another morainal area that was at least 65 years old. Between 1910, when the terminus hung above College Fiord, and 1935, Bryn Mawr retreated leaving a small, 1640 ft. long inlet. Around 1964, Bryn Mawr began to advance across this shallow inlet hanging above College Fiord. Between 1979 and 1986 the terminus advanced very slowly to near its 1910 position and is now approaching deep water. It would not be surprising to find Bryn Mawr beginning to retreat quite rapidly, especially if it pushes its protective terminal moraine over the edge of the hanging valley. Visitors might want to watch for an increase in the calving rate and the size of the bergs as these normally accompany retreat.

Smith Glacier

Location: Western side of College Fiord
Access: Boat

Type: Valley glacier from cirque basin
Distance from Whittier: 47 miles
Length: 6.2 miles (Field 1975)
Slope/aspect: Southeasterly
Status: Stable with minor fluctuations

Smith Glacier is the closest tidewater glacier to Harvard Glacier. Its twin tributaries originate in cirques 4000 to 5000 feet above the fiord. The glacier descends at a rate of 2200 feet per mile to sea level, falling in stairsteps through one spectacular icefall to the next. In 1910, Tarr and Martin estimated the height of its ice-cliff at about 100 feet. They found Smith's advancing margins most interesting: "Along the advancing margin the alders were being destroyed in three ways, — by actual overriding of the spreading glacier, by stream encroachment, and by ice-block avalanches which rolled some distance out into the forest, knocking down and breaking off shrubs and removing bark 6 or 8 feet above the ground (Tarr and Martin 1914, p. 301)."

After its advance during the first decades of this century, Smith retreated to near its 1899 position. During the early 1940s, it became more active and advanced, but by the late 1950s a slow retreat began followed by another period of weak advances and retreats.

Field reports that interesting enough sometimes Smith's ablation area is littered with massive rock debris originating from rock slides in its accumulation zone; othertimes the terminus is clear and free of rock debris. (Personal communication 9/16/87).

Fig-83. Harvard Glacier has been advancing for several centuries. Radcliffe is the first tributary on the bottom left. Station 1 on the small point in front of the glacier's eastern side is about to be overridden (1987). Photo by Austin Post, Aug. 24, 1964.

Harvard Glacier:

Location: Head of College Fiord
Distance from Whittier: 49 miles
Access: Boat
Type: Ice-calving valley glacier from Chugach Icefield
Length: 24 miles (Field 1975)

Area: 199 sq. miles (Post 1986, unpub.)
AAR: 80% (Post 1986, unpub.)
Slope: Southwesterly
Status: Between 1979 and 1986, Harvard advanced approximately 750 ft. along its eastern margin while remaining nearly static across the western half of the terminus.
Photos: Fig-83. Maps: Figs-76, 80, 84.

Recent History:

In 1909, Gilbert noted that Radcliffe Glacier barely joined Harvard and predicted that if Harvard receded moderately, Radcliffe would become a separate glacier. Instead of retreating, Harvard has advanced throughout this century. Part of this advance may be attributable to an increased flow from Radcliffe Glacier. However, the most important factor is that Harvard retreated to its retracted position sometime during the neoglacial period and has been advancing ever since.

Tarr and Martin (1914) noted that as Harvard advanced it pushed a moraine up to 15 ft. high along the western shoreline with a large peat-roll in front of it. In 1935, Viereck found a tree with 246 rings less than 500 ft. from the advancing ice-cliff which pushes back the date of Radcliffe and Harvard's last occupation of this area to at least the mid-seventeeth century. In fact, in this area where alder and willow communities dominate, trees are the exception rather than the rule. Tree-ring studies only indicate that the glacier has not occupied an area since the tree began to grow. They say little about how long it has been since a glacier retreated from the area. Between 1899 and 1986, Harvard Glacier advanced 5345 ft. The rate of advance appears to be a little over a mile a century.

Fig-84. Positions of the Terminus of Harvard Glacier: 1899 to 1986.
Sources: Field, *A.G.S. Glacier Studies.* Map No. 64-3-G2. Post, 1980e. Post, 1986 aerial photographs. For information on photographic stations cf. Fig- 80.

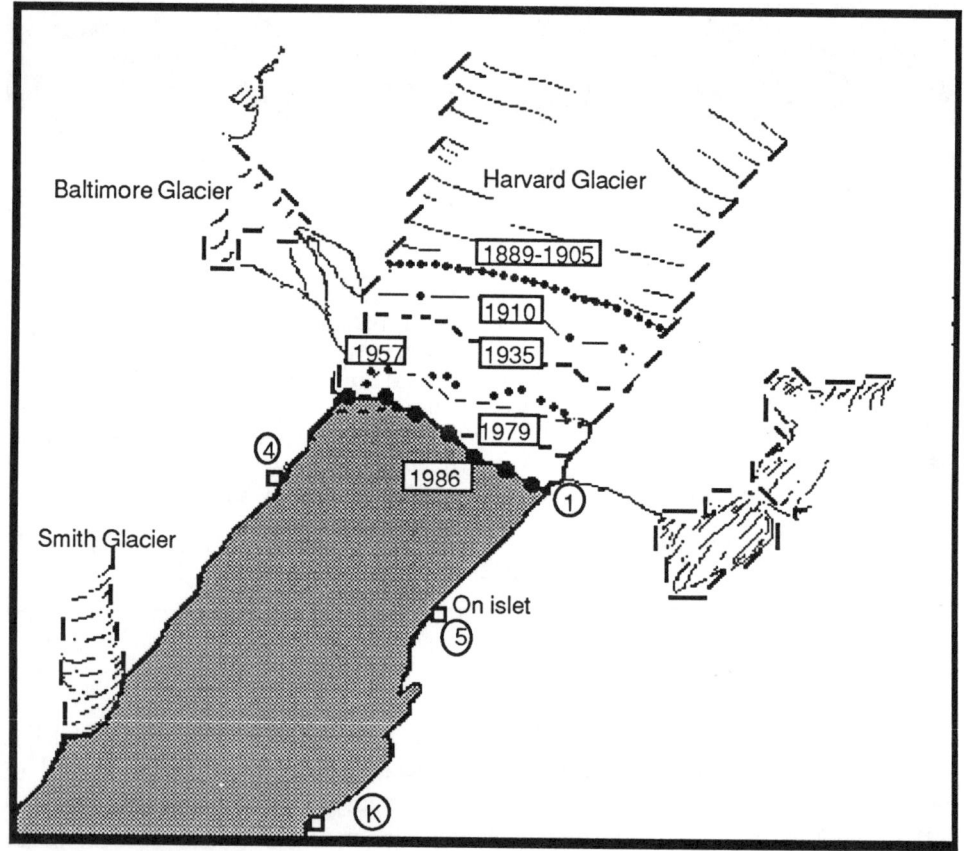

Chapter 11. A Tale of Two Glaciers: Meares and Columbia Glaciers

Visitors to Columbia and Meares glaciers have the unique opportunity to observe two glaciers, flowing from the same icefield that are in opposite phases of their cycles. After advancing for a 1000 years (and possibly as much as 3000 years), Columbia is now in catastrophic retreat, while Meares Glacier is continuing a slow advance that began following its own drastic retreat off the moraine that divides Unakwik Inlet (Heusser 1983) sometime around 500 years ago.

Meares Glacier:

Location: Head of Unakwik Inlet
Access: Boat,
Distance from Whittier: 56 miles
Distance from Valdez: 61 miles
Distance from Cordova: 83 miles
Type: Tidal, ice-calving valley glacier flowing from icefield

Length: 15.5 miles (Field 1975)
Area: 56 sq mi. (Post 1986, unpub.)
AAR: 85% (Post 1986, unpub.)
Aspect: Accumulation zone: SW;
Terminus: WSW
Status: 1979 to 1986: north side advanced approx. 700 ft; south side advanced about 100 ft.
Photos: Figs-22, 86-88
Map: Fig-85.

Recent History:

Meares Glacier was named by Grant and Higgins for John Meares an early English sailor who first arrived in Prince William Sound as a crew member under Captain Cook. Meares returned as captain of his own ship, the *Nootka*, which was engaged in the sea otter trade. Meares spent a miserable winter during 1786-1787 at Sunny Cove in Port Fidalgo (eastern Prince William Sound), where twenty-three of his men died of scurvy. It is unlikely that he ever visited Unakwik Inlet.

There is some disagreement over who first sighted Meares Glacier. Tarr and Martin following Davidson attributes this honor to Fidalgo. Davidson gives the following description of Fidalgo's account:

> Conducted by their Indian guides to the interior of this harbor (Puerto de Revilla Gigedo), they discovered in its depths to the north, a great level tract of snow which came to the water's edge, and ended at the base of the high mountains. Hardly had they seen this, when they

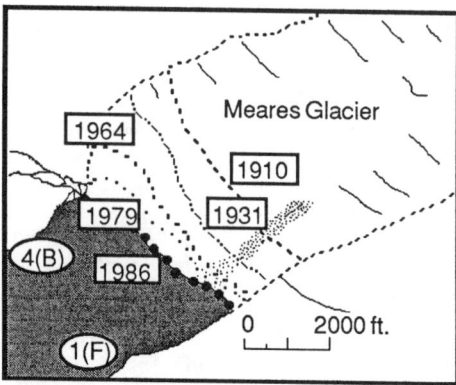

Fig-85. Positions of the Terminus of Meares Glacier from 1910-1986. Sources: Field, written communication; Post, aerial photos 1979, 1986.

Fig-86. The terminus of Meares Glacier in 1963. By 1986, Meares had advanced to where it diverted the course of Ranney Creek. A. Post photo. USGS.

Photographic Stations:

Photographic stations established by Grant and Higgins in 1909 and Tarr and Martin in 1910 can still be used. The best two are stations 4(B) and 1(F). The numbers indicate stations occupied by Field; the letters are cross-references to Grant and Higgins' and Tarr and Martin's designations. Station 4(B) is on a prominent rock ledge on the northern shore just west of the outwash delta for Ranney Creek. Station 1(F) is on a bulge of land protruding out from the southern shore which gives an unrestricted view of the entire face. Not pictured on the map are Sta. 5 which is on the prominent point near the bend and gives a good view of the northwestern end of the terminus and Sta. 6 on a rock ledge just above high tide on the outside of the dogleg's curve.

Field has a series of photographs of Brilliant Glacier taken from a boat which indicate that the glacier has been shrinking fairly steadily since he first photographed it in 1931. (Letter, March 21, 1986).

---

noticed that with each subterranean roar a mass of snow was thrown up from the center of the plain about half the size of the launch; fearful lest they should be overwhelmed or destroyed in this port, they did not continue their examination of that phenomenon, which is undoubtedly worthy the attention and investigation of a naturalist.
(Davidson 1904, p. 11.).

However, Whidbey, who had Fidalgo's maps and journal, states that Fidalgo did not go farther west than Long Bay (near Columbia Glacier). Most historians agree that Fidalgo's Puerto de Revilla Gigedo is Columbia Bay. Whidbey, himself, probably only crossed the mouth of Unakwik Inlet without exploring its inner recesses. Vancouver's summary of Whidbey's report gives distances in leagues as was his custom for distinguishing estimates from actual mileage. The section beyond the bend is not indicated on Vancouver's chart.

> . . . this arm was found to take a north direction, in general abut a league wide, and to terminate at a distance of about 4 leagues, at the foot of a continuation of the range of lofty mountains before mentioned. Its upper parts were much encumbered with ice, as were both the eastern and western sides with innumerable rocks, and some islets. (Vancouver 1798, p. 185)

The estimated distance of 4 leagues (12 miles) corresponds reasonably well with the position of the moraine (9 miles from Olsen Island). Whidbey is most likely describing winter pack-ice jammed up against the moraine during spring break-up or the mirage effect. From a distance, towering inferior mirages (cf. p. 80) often make the area north of the moraine appear to be completely blocked by ice.

In 1899, Capt. Glenn made a reconnaisance of the sound seeking a new and better route to the interior than Valdez Glacier or the Copper River. He appears to have been the first to round the bend in Unakwik. But because visibility was limited by rain and fog and heavy concentrations of ice barred his way, he described the dogleg as being divided into two arms: "When we arrived at the head of this inlet we found it divided into two arms, both of which were frozen over. The ice was largely slush, not strong enough to bear the weight of a man, yet we tried several times to break through with the bow of the steamer, which was very sharp, without sucess. We were unable, therefore, to get to the solid ice beyond the slush . . . . rain and fog prevented us from seeing very far into the mountains, but far enough for us to determine that no outlet exists at or near the head of this bay (Glenn 1899, p. 94)."

Capt. Abercrombie's 1899 map of Prince William Sound, however, correctly shows both the upper reaches of Unakwik Inlet and the glacier.

Grant and Higgins (1905) were the first scientists to see Meares Glacier. In 1909, they returned to map the terminus, describe and photograph the glacier, writ-

Fig-87. The ice cliff at the face of Meares Glacier towers above the sailboat. The hanging glacier in the background separated from Meares shortly after 1935. Photo by Nancy Simmerman, 1983.

ing: ". . .the Meares Glacier, although not so large as others, is one of the most beautiful ice streams of Prince William Sound. The end of the glacier is about 0.8 of a mile wide and forms a vertical wall of pure ice which is estimated to be 300 ft. high. It is actively discharging. The glacier is formed by two ice streams which descend from lofty mountains" (Grant and Higgins 1913, p. 25).

Meares Glacier still positively glistens. Its active ice has almost no ablation moraine; it is severely crevassed as far as the eye can see; a strong lateral moraine runs along the northern side; and a weak medial moraine marks where its two visible tributaries join. The ice-cliff discharges so much ice that it is often difficult to get through the brash that collects between the morainal bar and the dog-leg to the north.

On the basis of the vegetation in front of Meares Glacier, Grant and Higgins felt that the glacier had probably not been farther forward in two to three hundred years. This was confirmed by Viereck. He found a 280 year old tree near the ice-front (Viereck 1967, p. 199.)

Meares Glacier has been more or less steadily advancing since observations began. The north margin is advancing fast enough to move into live alders and spruce forests before the cold layer of air surrounding the glacier kills them. On the south side, one can see a pronounced barren zone during the summer months. This may be caused either by an annual advance and retreat of the glacier or by frost-chill.

According to Field, "one of the best local landmarks to measure any change in the position of the terminus is Ranney Creek. In 1974, the terminus was reported over the creek, but it had retreated slightly by 1976 (Field, letter, 9/16/87)." In the fall of 1986, the northern edge of Meares' ice-front was pushing into Ranney Creek's channel, diverting the stream.

The smaller valley and apron glaciers of Brilliant, Ranney and Baby lie in the upper half of Unakwik Inlet. The presence of mature vegetative communities indicates that these glaciers have not advanced in several centuries, while bare rock around them points to their continued slow retreat.

Fig-88. Brilliant Glacier is one of several smaller glaciers at the head of Unakwik Inlet that have been retreating slowly. Photo 1984 by author.

Columbia Glacier:

Location: Head of Columbia Bay
Distance from Valdez: 27 miles
Access: Charter, tour boats, ferry boat from Valdez, Whittier, and Cordova; cruise ship; flightseeing by plane or helicopter from Valdez, Cordova, Anchorage.
Type: Ice-calving valley glacier flowing from icefield

Length: 42 miles (Post 1986)
Area: 435 square miles (Post 1986)
AAR: 67% (Post 1986)
Aspect: Accumulation area: SSW; Terminus: S
Status: Drastic retreat. Between 1981 and Sept. 1986, Columbia retreated 3500 ft. on the east side, 8000 ft. in the center, and 2000 ft. along the west margin.
Photos: Figs-8, 14, 16, 24, 25, 36, 90-94
Map: Fig-95

Glacial History: The Pleistocene and Holocene periods.

The smooth, rounded ridgetops surrounding Columbia Bay are indicative of glacially overriden areas. Both Heather and Glacier Island exhibit typical roche moutonnée silhouettes. Smaller glaciers scoured out cirque-shaped bays and submerged hanging valleys on the mainland and Glacier Island.

Radiocarbon dates taken from a layer of duff adjacent to bedrock at the northern end of Heather Island indicate that this area was ice-free by about 9300 years ago (Post 1975). In 1974, Columbia Glacier advanced to within 1640 ft. of this site. During its 1917-1922 advance, Columbia came to within 492 feet of it.

Little is known about Columbia's activity during preceding neoglacial phases, but it undoubtedly occupied a position much farther up the fiord. As Columbia wastes, it discharges logs from overriden forests or uncovers duff layers whose radiocarbon dates provide useful information regarding the area's past. To date, Post has obtained the radio-

Neoglacial Age Radiocarbon Datings from Columbia Glacier	
Location	Years Before Present
West side of Tropic Island	less than 200
Bare cliff on west side of Columbia Bay near recent terminus	less than 200
Nanatuk Lake (below waterline)	130 + or - 150
Tenas Nanatuk (below waterline)	300 + or - 145
Tropic Island, lower till	395 + or - 150
Upper edge of Terentiev Lake	825 + or - 55
Iceberg rafted wood	840 + or - 125
Kadin Lake	840 + or - 80
Terentiev Lake Basin (below waterline)	1650 + or - 150

Fig-89. Based on unpublished work provided courtesy of Austin Post.

carbon dates (personal communication) shown in Fig 89. As expected, radiocarbon dates for the western cliff, Tropic Island and Nanatuk lakes areas indicate that Columbia advanced into these regions relatively recently, possibly since Cook and Vancouver's first visit or perhaps a century earlier. The Nanatuk lakes were formed when the glacier blocked off their drainage area. Tree tops still protrude above the surface of the lakes.

The oldest dates recorded so far suggest Columbia moved into the Terentiev Lake basin around 300 AD (cf. Fig-36). Dates from the upper elevations of Terentiev and Kadin Lakes seem to indicate that the glacier gradually deepened backing up both ice-dammed lakes behind its terminus until these areas were submerged sometime around the twelfth century. As Columbia retreats and more radiocarbon dates are obtained, glaciologists hope it will be possible to fill in the history of its advance more completely.

Glacial History: Discovery, Observation, and Changes

During the age of exploration in Prince William Sound, several explorers noted the glacier. Cook, who anchored at Snug Corner Cove, was the first European to enter the sound, but he failed to mention nearby Columbia Glacier or any ice. The skilled, Spanish

Fig-90: When the Harriman Alaska Expedition visited Columbia Glacier in June of 1899, Gilbert established a photographic station, Gem, on this point near the western terminus. The photograph shows that Columbia Glacier had recently shrunk. Stations Muffin, Bonsai and BoomBoom are on top of the first rise on the left. Photo by G.K. Gilbert, No. 355. June 1899. World Data Center Archives.

explorer Lt. Salvador Fidalgo sailed between Glacier Island and Point Freemantle to near Long Bay. Fidalgo reported a volcano in this vicinity on his maps after misinterpreting the loud booming noises from the glacier's calving activity for vulcanism. Both Whidbey and Johnston reported Columbia Glacier to Capt. Vancouver (1794).

Russian fur traders and explorers also visited the area. Capt. M.D. Teben'kov, who later became governor of Russian Alaska, felt it necessary to discount Fidalgo's volcano when he wrote: "In 1848, I saw them (Chugach Mountains) in all their grandeur and did not notice any volcanoes (Teben'kov 1852, p. 21)." (Post later named Kadin and Terentiev Lakes after Teben'kov's two cartographers.)

At this time, Columbia Bay was variously called Glacier or Freemantle Bay. In 1899, the Harriman Alaska Expedition stopped here and renamed it Columbia Glacier for Columbia University. By studying photographs taken by the Harriman Alaska Expedition, glaciologists have been able to record the changes in the glacier since 1899.

During most of this century, Columbia Glacier's terminus exhibited minor fluctuations, making a few small advances (1898, 1910, 1922, 1935, 1974) followed by retreats. Moraines and vegetative evidence for these advances occur along both shores and on Heather Island. Its recent advances have not crossed the 1920 and 1935 push moraines.

Fig-91. In 1909, Grant and Higgins found Columbia Glacier advancing and knocking trees over on Heather Island. The rock and tree in foreground can still be located. Photo: U.S. Grant, No. 87. June 24, 1909. World Data Center Archives.

Fig-92. In 1976, Columbia Glacier was still riding up on Heather Island. A large embayment had formed in its Columbia Bay side (left side). Glacier dammed lakes Terentiev and Kadin are in the valleys on the left side of the photo. Station Boomboom is atop the first ridge on the left and overlooks Terentiev Lake. The Nanatuk lakes are a chain of lakes on the right hand (eastern) side. The Great Nunatak is in the middle. Photo by A.Post, USGS, 1976.

Shrinkage in Columbia Glacier's height and its failure to eliminate embayments caused some glaciologists to suspect that dramatic retreat might be imminent. Post and Meier (1980) predicted Columbia would back off its moraine and enter the drastic retreat phase of its cycle in 1982 or 1983. In December 1984, USGS scientists at the glaciology office in Tacoma announced that irreversible retreat was well underway. Compared to a glacier's slow advance (approx. 1 miles a century), retreat has been rapid. Aerial photographs taken on June 13, 1986 and again on September 9, 1986 showed that the central portion in a depth of about 1200 feet of water retreated nearly 3000 feet in a mere 58 days. By August 22, 1987, Columbia Glacier had retreated 2 miles (10,050 ft.) from its terminal moraine (personal communication, A. Post).

Massive icebergs possibly weighing as much as a million tons will continue the glacier's sculpturing work by gouging and reshaping the terminal moraine, shoal areas, and beaches. Navigational charts may not be accurate.

The retreat has caused quite noticeable changes in the face of the glacier. Formerly, as Columbia pushed up on to its moraine, ice-cliffs rose towering into the sky. Great seracs tilted precariously for days before crashing into the sea. Now, without the protection of the moraine, warm sea water undercuts the glacier's face and penetrates into crevasses. Huge bottom bergs like fabled monsters of the deep may suddenly surge to the surface from depths near a 1000 feet. The ice-cliffs have been replaced by a gently sloping face.

In place of its former dramatic ice-calving events, Columbia now has an aura of the mysterious. What lies concealed under its blanket of ice? One islet emerged during the summer of 1986 just off the western shore. Will the new fiord be straight and clear? Will it have more islands? Will it have narrow channels and hidden bays?

The Glaciers of Prince William Sound, Alaska

Fig-93. By 1981, Columbia Glacier had retreated well back from Heather Island. A large embayment called "The Ballroom" had formed opposite Heather Island. The eastern point of the Ballroom is on Tropic Island. A portion of the glacier still remained on the terminal moraine in Columbia Bay. However, barren areas on the hillside by the western margin show that Columbia was shrinking. Glacier-dammed Lake Terentiev lies in the first valley on the left. Kadin is in the second. Photo by Austin Post, USGS, Sept. 26, 1981.

Fig-94. Columbia Glacier is in drastic retreat. The moraine and Heather Island keep much of the ice from flowing out into Columbia Bay. Tropic Island is now ice free. Dark areas by the glacier mark the mouths of glacial streams. Stations Echo and Top are on the point to the right of Tropic Island. Photo by Austin Post, USGS. Sept. 12, 1986.

An Observer's Guide to

Fig-95. Visitors photograph Columbia Glacier from Photographic Station 3A. This station is atop the 1957 push moraine. Photo by Nancy Simmerman, 1984.

Tourists and Prospectors:

Columbia Glacier has been the most frequently visited of Prince William Sound's many tidewater glaciers. Since the Harriman Alaska Expedition brought the first tourists to see it, thousands of men and women have traveled to Alaska not for its furs or gold, but to watch huge blocks of ice crash from Columbia's impressive four mile wide face. Its economic importance to tourism based in Valdez (and to a lesser extent in Cordova and Whittier) has been tremendous. In 1985, one tourboat operator carried 14,000 guests to view Columbia Glacier. Over the years, it has been the scenic backdrop for advertisements, home movies, and Hollywood and Japanese films. Now, with its drastic retreat, Columbia is a star in its own right.

The retreat might not be newsworthy except for two facts. First is the scientifically significant fact that Columbia is one of the last remaining North American iceberg-calving glaciers to undergo a drastic retreat from its extended neoglacial position. Second, the location near the oil tanker shipping lanes means any increased discharge of icebergs the size of bergy bits could possibly pose a hazard to shipping.

For those who wish to spend some time in the area, the best anchorage is at Dalli Cove on the eastern side of Heather Island. From the anchorage it is an easy walk to the northern end of Heather Island. This area will soon be too far from the glacier for good viewing but it will remain historically interesting for its old moraines and photographic stations.

Those interested in the revegetation of a glaciated area will find this area particularly intriguing. Near the glacier, river beauty *(Epilobium latifolium)*, which spreads by lightweight, wind-born seeds and is very tolerant of heavily mineralized soils, sunlight and wind, is an early colonizer. Mosses, lichens, mushrooms, alders, willows, and herbaceous plants such as hawkweeds *(Hieracium triste, H. gracile)*, yellow-flowered willow herb *(Epilobium luteum)*, Horn's fireweed *(E. Hornemannii)*, and single-sided wintergreens *(Pyrola secunda)* may take decades to become established. Still they are early colonizers compared to conifer forests. Fungi play a very important role in making nitrogen available to plants. Without nitrogen pioneering spruce and hemlocks grow only to several feet in height before turning yellow and dying.

Revegetation is occurring much faster at Columbia than at Nellie Juan. Several factors may be significant. First, the prevailing winds at Columbia carry seeds towards the glacier, whereas at Nellie Juan they carry them away. Secondly, these same winds are much colder and harsher blowing down off Nellie Juan than the warm, oceanic winds that blow towards Columbia. Wind chill may play an important role in inhibiting rapid revegetation at Nellie Juan. The environs around Columbia seem much warmer and more hospitable. Third, sedimentary rocks provide the mineral content in the developing soils at Columbia, while more acidic granitic rocks form the major mineral constituent at Nellie Juan. Finally, in some areas rocks carried down the surface of Columbia Glacier are already colonized by mosses and lichens. At Nellie Juan, the topography constricts the glacier's terminus to a narrow gorge where glacial debris is deposited into the ocean.

In 1983, it was possible to take a small boat over the moraine at the head of Heather Bay to No. One River; increased discharge of brash ice, growlers and bergy bits made this more difficult by 1985. Great care should be taken in crossing the moraine since its depth is still being modified by the addition of ice-rafted boulders and sediments. The eastern side of Columbia Glacier is a fascinating place to explore. One can see the 1910, 1922, 1935 and 1974 push moraines, old trimlines, logs and trees from an overriden forests, and eskers. In 1984, it was still a relatively easy walk across the gently sloping eastern edge of the glacier to the Great Nunatak or up the valley to the Nanatuk Lakes (cf. p. 123).

Columbia provided prospectors with an access route to the peaks and ridges of the Chugach Mountains. Supplies for mines such the Cameron-Johnson, Gold King, Rough and Tough, and Mayfield Mines were freighted up the glacier's east side. During the early part of this century, several warehouses belonging to the miners and freighters were located along the shores of Columbia Bay. With the advent of aviation, some of the mines were reopened. Bob Reeves made a name for himself as the "Glacier Pilot" by flying supplies to mines such as his own Rough and Tough Mine. None of the mines ever reported much production. For example, over about 40 years the Rough and Tough Mine only reported production of 72 ounces of gold and 20 ounces of silver.

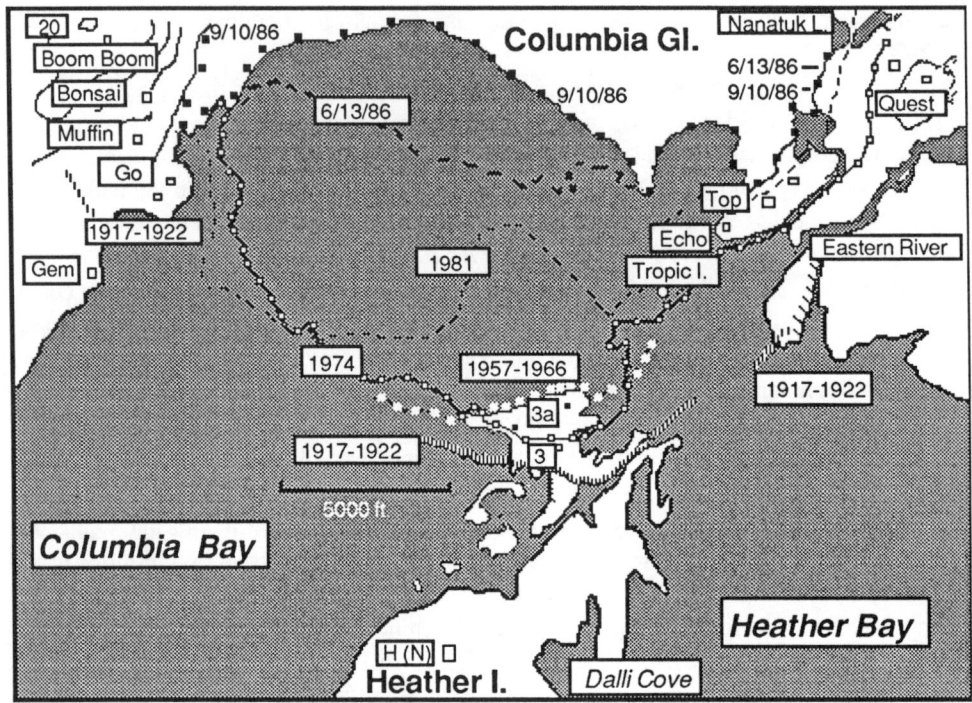

Fig-96. Positions of Columbia Glacier's terminus.

Sources:

W.O. Field, unpublished maps "Columbia Eastern Land Terminus, 1964," "Columbia Main Ice Front, 1966."
A. Post, *Preliminary hydrography and historic terminal changes of Columbia Glacier, Alaska* (1975).
A. Post, USGS aerial photos June 13, 1986 and September 10, 1986.
Place names are from A. Post or USGS map.

Photographic Stations:

Various scientific parties have established numerous photographic stations around the terminus of Columbia Glacier. The following are a few of the most important ones.

The northern summit of Heather Island (Station H [N]) provides a sweeping view of Columbia Glacier that can be compared with photographs taken by the Harriman Alaska Expedition (1898), and Field (1931, 1935, and 1964). We have bushwhacked up from Dalli Cove. Field recommends going directly up the west side from Columbia Bay.

Station 3 is accessible from Dalli Cove by walking through a series of peatland bogs, around a lake, and across a tidal flat (the northern shore of which is the 1917-1922 push moraine) to a knoll on the northern tip of the island. Although now a bit far from the terminus, this is still an excellent station for viewing push moraines left by Columbia's activity during this century. A series of push moraines from earlier advances can still be detected. Push moraines from advances made in 1917-1922 are in the woods behind the station. The 1935 moraine is located immediately below the station. The 1931 moraine passes just north of 1935. The 1957, lies north of 1931 on a ridge marked by the cairn for station 3A. Finally, the 1964 push moraine is down the ridge from station 3A. These are the most prominent push moraines, many others crisscross the area.

From Dalli Cove one can take a skiff to the eastern land terminus or with considerable caution one can take a boat across the moraine and anchor off Eastern River. Station Tropic is on the highest point of Tropic Island and is marked by a cairn. Station Echo is on the point to the east and is marked by a cairn with a protruding stick. A ten minute walk up the ridge from Echo is Station Top near some dead trees that were overridden by the glacier. The glacier stripped the bark from the upper surface but left it on their lower surface. Station Quest and Nunatak are reached from Eastern River and are atop the most conspicuous knoll. Post says that Quest is a superb, permanent viewpoint.

Near the shore on the western side of Columbia Bay there are three good photographic stations. Station Gem is on a bare, rocky point and marked by a conspicuous pipe. Since Gilbert established it in 1899, it has been reoccupied repeatedly. In 1986, Post established two new photographic stations on the recently vacated moraine. One is located by a large, block-shaped boulder, the other a little farther north atop the moraine.

For the more adventuresome, four stations at higher elevations provide outstanding vantage points from the western side. Post considers Station Muffin in a bog atop the 600 ft. knoll to be good, but recommends Station Bonsai as superb. Muffin is reached by heading up a game trail in the draw between Station Gem and the moraine. To find Bonsai contour north around the knoll from Muffin. One comes to an opening which looks like it ought to be a station except that the view is blocked by trees. About 400 ft. through the trees from here, there is another opening backed by a large rock about 20 ft. high. Go around the rock and climb to the top. This is Bonsai. A series of panoramic photographs taken from Bonsai by Post show an unobstructed view of the entire terminus and for many miles up the glacier.

Field did not use Station Muffin or Bonsai but hiked to Station 19 (Boom Boom), which is the top of the 1650 ft. ridge. This can be reached by continuing through the openings and up the ridgeline from Station Bonsai. He considers Station 20, on the lower summit about approximately 1600 ft farther north, an even better viewpoint, describing it thus: "It's a fairly laborious hike from the shore, but if one has the time on a good day it is certainly rewarding. It also provides an excellent view of the ice-dammed lake (Terentiev) to the west. Once on the summit ridge, which is above timber-line, travel is easy. . . For a full sweep of the lower few kilometers of the glacier it has no equal." (Letter to author, March 21, 1986).

As the glacier retreats, new stations will be established. Two stations have already been placed on the Great Nunatuk by Post. Station Susie is on the first outcropping just north of the south finger of Columbia Glacier. This finger separated from Columbia Glacier in 1987 making the Great Nunatak no longer a nunatak. Station Grand Central is located on the northwest side at about 100 ft. elevation and commands a good view of the upper glacier. Formerly, it was possible to walk from the Nanatak Lakes along the east margin of the glacier to the Great Nunatak, but in 1987 a stream from one of the Nanatuk Lakes began flowing along the east margin of the glacier making access to the glacier difficult or impossible (personal communication, A. Post).

Chaper 13. On the Trail of the Prospectors: Glaciers of the Port Valdez Area

Port Valdez, an east-west tending fiord, lies in the extreme northeastern corner of Prince William Sound. During the late Pleistocene ice ages, a large glacier reaching about 3200 ft. above sea level flowed down Port Valdez, squeezed through the narrows and merged with the vast piedmont glacier covering Prince William Sound. No recessional moraines from this glacier have yet been identified. Radiocarbon dates from peat samples taken from test borings at the new city of Valdez site (the former was abandoned following the 1964 Earthquake) indicate that Pleistocene glaciation ended in Port Valdez sometime before 9,520 years ago (Williams and Coulter 1979).

Recessional moraines belonging to tributary glaciers that advanced during the Holocene neoglacial periods have been found near the southeast corner of the Alyeska terminal (Allison Creek area). Striation marks are visible on tidal bedrock near the Fort Liscum

Fig-97. The town of Valdez sprang up in 1898 as thousands of prospectors arrived here in hopes of crossing Valdez Glacier and reaching the Klondike gold fields. Valdez Glacier looms in the foreground. Camica Glacier, which was a tributary until 1910, hangs in the valley behind. Photo by Cameron, circa 1910. Mary Starr Collection. Courtesy of the Valdez Heritage Center.

Fig-98. Until the opening of the Military Trail in 1899, the major route for prospectors heading from Valdez to the interior gold fields went 15 miles up Valdez Glacier before crossing a pass leading to the Klutina Glacier. Photo by Post, Aug. 24, 1964. USGS.

Landslide which is just east of Allison Creek. A neoglacial moraine belonging to an earlier advance of Valdez Glacier lies along the eastern side of its valley.

With the rise in sea-level about 8,600 years ago following the melting of the Wisconsin ice-sheets, the coastal freshwater bogs that underlay the site of new Valdez flooded. Subsequently, the area was slowly reclaimed from the sea as marine sediments and outwash debris from Mineral Creek suceeded in constructing a typical fan-shaped delta behind the bedrock islands (Fig-40).

Today, the fiord's overly steepened sidewalls, hanging U-shaped valleys, cirques, spectacular arêtes and lovely, horned-shaped Sugarloaf Mountain are scenic reminders of the area's past glaciation.

Recent glacial activity is confined to two large glaciers, Valdez and Shoup, and to many smaller cirque and apron glaciers. From the small boat harbor, one can see thirteen cirque and apron type glaciers, some of which extend into small valleys. Anderson Glacier is the prominent glacier draping the mountainside between Sawmill Bay and Shoup Glacier. It is slowly retreating.

Valdez Glacier:

Location: Head of Port Valdez
Access: 7.4 miles on the Richardson Highway from downtown Valdez. Take the airport turn-off and drive 3.5 miles to the end of the dirt road.
Type: Valley glacier descending from the Chugach Icefield
Length: 21 miles (Field 1975)
Area: 61 sq mi. (Field 1975)
Slope/aspect: Accumulation area, south; terminus southsouthwest
Status: Retreating
Photos: Figs-20, 21, 97-99
Map: Fig-100

Glacial History:

Valdez Glacier flows from the Chugach Icefield that feeds all the glaciers along the northern part of Prince William Sound. It probably has not extended to tidewater since the end of the Pleistocene. During this century, Valdez Glacier has been steadily retreating except for spasmodic advances early in this century. In 1910, Tarr and Martin noted that Valdez Glacier calved into several small lakes, but no large lake existed. In 1935, Field found that continued shrinkage had lowered its height 15-20 ft. a year which left the Ramsey-Rutherford Mine, originally built on a level with the ice, stranded 200 feet up a precipitous cliff. During most of this century, Valdez Glacier has been retreating at an average yearly rate of 45 to 55 ft (Field 1975). The only known advance occurred in 1906-07 when the glacier advanced 250-350 feet according to the records of Dr. Camica (Grant and Higgins 1913). This advance was most probably caused by a kinematic wave.

Tarr and Martin (1914) photographed the glacially-dammed lake between Valdez and Camica glaciers. Icebergs stranded along the shoreline indicate the lake had recently dumped. Local reports suggest that this lake formed every spring and emptied into a tunnel beneath Valdez Glacier. As Valdez Glacier retreated, a 200 ft. wide outlet channel openned, and the lake no longer forms. However, periodically icebergs block this channel causing the lake to reform temporarily. If Valdez Glacier should experience another kinematic wave advance, such as the 250 to 350 ft. advance in 1906-1907, it could block the

Fig-99. Valdez Glacier has been retreating for most of this century. In 1910, the glacier still extended well around the bend. The terminal lake seen in Figs-98 and 100 had not yet formed. Cantwell, circa 1910. Simon Poot Collection. Valdez Heritage Center.

channel again and cause the lake to reform (Post and Mayo 1971, Rundquist 1981).

Valdez Glacier has three other glacially-dammed lakes (Post and Mayo 1970, Rundquist 1981). The most important one is Rundquist's Lake No. 3 (No. 24, Post and Mayo), which lies in the valley between Prospectors Peak and Abercrombie Mountain about five miles up the glacier. At present, Lake No. 3's outbursts occur subglacially at erratic intervals and do not have an exceptionally large peak discharge. However, as the ice-front retreats, Lake No. 3 will grow in volume, and its outbursts could cause increased downstream flooding. The best way to see this and other glacier-dammed lakes is by a flightseeing trip from Valdez.

The Valdez Glacier Route:

Valdez Glacier has played an important and exciting role as a transportation corridor connecting the coast to interior Alaska. Today, the Richardson Highway runs through Keystone Canyon and over Thompson Pass linking Valdez to the interior. In 1899 the Army opened up a narrow trail, known as the Military Trail through Keystone Canyon. Prior to this the canyon was considered an impassible gorge made treacherous by its precipitous, wet and slimy moss-covered walls and stupendous waterfalls. There was no trail yet from Seward or Anchorage to interior Alaska and the Copper River route was impassible in the summer. Because of its low-gradient and relatively smooth surface, the glacier provided a route from the sea port of Valdez to the interior.

In 1884, Lt. W.R. Abercrombie was charged with finding a route from the coast to the interior. After spending almost three months attempting to ascend the Copper River, Abercrombie hired "the son of Plutoniff" to guide him across Valdez Glacier. This would seem to indicate that the man knew the route. The presence of Russian traders in the interior are also reported in Copper River Joe's account *The Golden Cross (?) — on the Trails from Valdez Glacier* (pp. 75-76). Lt. Abercrombie relates the recent history of the route in his 1884 account of his explorations:

> Its history related by the old Russian, is as follows: Some years ago, probably twenty or thirty, the portage was used entirely by the Upper River Indians, who came down the Copper River to the stream heading in the lake, which not being previously named or visited by white men, is designated Lake Margaret. Up this they traveled to the lake, hence to the foot of the passage. Here they left their bidarra and packed their furs over to salt water, where bidarras were furnished by the Chigachimutes for the voyage to Port Etches. For this service the Upper River Indians paid a tribute in furs, not only for the bidarra, but also for the privelege of passing through the country.
>
> On one of their periodical tours there arose some disagreement, and words led to blows. The result was the annihilation of the Copper River party by the Chugachimutes. This occurred in the spring. In the fall an epidemic of some kind carried off over half the inhabitants of the village, which caused the remainder to flee not only from the epidemic (which they attributed to the powers of the shaman of the Copper River natives), but from their wrath . . . After the fight . . . the Upper River Indians adopted the route via the mouth of Copper River.
>
> (Abercrombie 1884, p. 391)

The Abercrombie party was the first American expedition to cross Valdez Glacier to Klutina Lake. He left an account of this crossing that tells more about his pride and stamina than it does about the glacier:

> On the morning of Sept. 12 the expedition started up the trail at daylight, two half-breeds in the lead. Each had a blanket and a dried fish. Myself and Brumback carried a blanket and overcoat, rolled up in which was some frying-pan bread for supper and for breakfast to partake of on the lake. After an arduous climb for six hours we reached the glacier in the following order, viz. half-breed Russian, son of Plutonif, myself, Nicholi Necolsky, half-breed from Nuchek, and Lieutenant Brumback. After travelling an hour or more the fog settled down and we could not

> see more than 50 yards. About this time a call from the rear was heard from Lt. Brumback . . . . it was found that he had taken with violent cramps in the muscles of his legs. On returning to him he was found lying in the snow. I agreed to go far enough to see the lake and locate its outlet and return to him. The snow was very soft and rotten and the guide fell through. Had I not been close at hand to drag him out, the guide would have lost his life. No one could see the bottom of the fissure, but water could be heard rushing past in the darkness below. After locating the lake and its outlet, I returned and found Lieutenant Brumback still on the snow. The guides informed me that the snow was too thin for packing, but that in two months the passage could be made safely. On returning to camp, which was reached after dark, I found my feet partly frozen. . . (Abercrombie 1884, 391-392).

Abercrombie is describing a day's hike of over 40 miles through heavy brush in mountainous terrain and across a glacier. He reported that "The estimated altitude of the highest point of this portage is about 2,500 feet, and from the base of the mountains on the west to the lake is about 15 miles." (Abercrombie 1884, 392). Abercrombie's understatement was to be bemoaned by many a prospector.

In 1896-1897 after gold was discovered in the Klondike, thousands of men left from west coast ports for the Alaska gold fields. Finding a route from the coast to the interior became a major problem. On arriving in Prince William Sound, cannery personal at Orca (near modern Cordova) advised the prospectors to try Abercrombie's route over Valdez Glacier. On April 18, 1898, the Copper River Exploration Party, whose members were from the Army and U.S. Coast and Geodetic Survey, arrived in Valdez. Prospectors angrily informed them that more than a thousand people, under the impression that the glacier led to a lake in a short distance, had been toiling up Valdez Glacier for over a month with no indication yet of what lay on the other side (Schrader 1900, 350). Schrader (USGS) and Lt. Brookfield were sent to make a hasty reconnaissance. On their second attempt, they returned after three days to report that the summit was about 4800 feet instead of the rumored 1500. And, they related to those at the base of the climb that already more than 2000 people had crossed over into the Klutina.

The route over Valdez Glacier at best was hazardous and arduous. It was also one of the safest and fastest routes to the interior. Three to five thousand gold-seekers crossed Valdez Glacier during the year 1898-1899. A few of these prospectors left accounts of their experiences. George C. Hazelet describes the glacier in his diary:

> The glacier is a wonderful thing. It begins almost at the edge of the sound at the time of year we reached it [March], and extends in a general northerly direction for about 15 miles to its summit. Here it reaches a height of 5140 feet as indicated on our barometer, and crosses over the range [where] it runs down on the north slope of the coast for a distance of 10 miles. On either slope it is made up of a series of

benches, having on the south slope about seven in all, six of which one can transport goods up easier by windlass or rope and pulleys better than any other way. The man that starts to pull 1500 lbs. of supplies from Valdez to the top of the summit needs plenty of grit . . . .

The trip from the top of the summit to timber on the north side is not so bad. Yet there are some four benches to go down. One simply loads his sled with 600 - 800 lbs. of goods, binds it so it cannot slip as he thinks and starts off. All goes well till your sled gets the upper hand of you and is pushing you downhill at breakneck speed, when you are compelled to run it off the track and into deep snow to stop it . . . .

Icefalls divided the glacier into seven benches. At the foot of these benches, tent-towns of several hundred people sprang up. Most parties expected to transport their supplies over the mountains in two weeks; most took five to six weeks. Not only was the work physically exhausting, the miner also faced the quadruple hazards of glacier travel — hidden crevasses and thundering avalanches, sudden blizzards and debilitating snow blindness. Hazelet records his own near misstep into a crevasse:

I was working away a few days ago, thinking I was on solid ice, and put more weight on one foot than the other. The snow gave away and I looked down and there was a great hole. Three of us tried for 15 minutes to fill that hole with snow and could not. It must have been 300 feet deep. A man would stand but little chance in a place like that (March 29, 1898).

Blizzards and avalanches brought the prospectors both an obligatory rest from their labor and an understandable sense of frustration at the additional delay. And yet, men like Hazelet also took time to appreciate their experiences.

It has stormed all day, rain changed to snow and a very heavy wet snow has fallen. This PM the snow has been sliding down the mountain at a great rate. It sounds like distant thunder as it slides down pulling rock and dirt with it to load the old glacier heavier and still heavier. All of these sights are wonderful to the person born and raised on the level prairies of Iowa and Nebraska (March 30, 1898).

Hazelet was also a keen observer of those traversing the glacier with him:

At present writing there are about 2200 people on this trail. Goods are scattered from one end of it to the other. There are people of all nationalities but I think the Swedes predominate. There is the long lean man, the short fat one, who puffs and blows like a porpoise as he tugs at his load up the glacier . . . How I pity some of these people, tired, footsore and weary. They trudge along looking as if they could go no further. I think many will give up and drift back to where they came from. . . .

As Hazelet predicted many of the gold-seekers were poorly prepared for the trials ahead — a wilderness with no towns, no public transportation, no rescue or medical services, no established system of communication, and no stores from which to purchase supplies. More than half returned across the glacier to winter, destitute in Valdez. The Valdez City minutes for 1898 make brief mentions of the dangers: "Nov. 19th —People began to return to Valdez from the Interior. (Illegible) died on the summit and was brought down the following week at the expense of the town. Henry Krohn was frozen on the glacier Nov. 20th and died in town on Nov. 29th." Contributions were made by town citizens and businesses to build a relief station on the Glacier. The Whaling Company donated lumber. The government hauled the materials up the glacier at no charge. The station was equipped with a tent, sleeping bag, blanket and stove. Melvin Dempsey who had started a Christian relief station on the beach also contributed his labor and supplies to the glacier relief station (reprinted in *Valdez Vanguard,* July 2, 1980).

The Valdez Glacier route was mostly abandoned in 1899 following the cutting of a treacherous, but passible trail through Keystone Canyon. Freighters continued to use Valdez Glacier for several more years until the Military Trail was widened into a road. Between 1912 and 1928, following the discovery of gold on the mountainside above Valdez Glacier, tons upon tons of freight including an entire stamp mill were hauled by dog sled to the Ramsey-Rutherford Mine. In the 1920s, trappers in the Lake Klutina area regularly crossed Valdez Glacier. One party snowshoed the 35 miles in one day. A tragedy occurred on Valdez Glacier in 1943, when a group of 21 Army Signal Corps Men encountered a blizzard. One man died in a crevasse and the rest required hospitalization for frostbite. Several underwent the ordeal of amputations; two remained hospitalized for over a year.

On a happier note, in 1959 a group of experienced mountaineers led by glaciologist Lawrence Nielsen combined skill and modern technology to explore the glacier and climb its upper peaks. After flying into a base camp at the summit, the party climbed Mt. Brookfield, Mt. Mahlo, Townsend Peak, Mt. Abercrombie, and Prospectors Peak. They then hiked down the glacier to Valdez. Nielsen's comment on their downhill trek tells more about the feats of the '98ers than many a prospector's laconic diary entry. "We were doing things the easy way compared to the 1898 prospectors; our equipment was airlifted twenty-five miles from Valdez and deposited at an elevation of five thousand feet. Even then we found it no picnic to lug the equipment back downhill to Valdez, but we certainly were thankful it was not the other way! (1960, 33)."

Points of Interest:

To reach Valdez Glacier take the airport turn-off from the Richardson Highway. It is 3.5 miles from the turn-off to the end of the road at Valdez Glacier Lake. The road crosses Valdez Glacier's outwash plain; formerly, the outwash plain flooded frequently as a result of heavy rainfall and the seasonal dumping of one or more of the glacially-dammed lakes. Repeated diking and the glacier's retreat have reduced the flooding and allowed a riparian woodland of sinuous alder and graceful cottonwoods to grow along the waterway. Near the lake in front of Valdez Glacier the outwash plain is mostly barren; the old moraines and kettles, and newer alder and black cottonwoods have vanished under the heavy equipment of a sand and gravel company. In this area, blasting has removed most traces of the lateral moraine and striation marks on the rocks.

Today, Valdez Glacier, itself, is almost invisible from the gravel pit area adjacent to the lake at the end of the road. Only a small portion of the terminus, heavily darkened by ablation moraine, can be seen aground in the lake. The rest, which extends 14 miles to its cirque basin, is obscured by a point.

Moraine-covered icebergs, adrift and aground in the lake, provide excellent examples of how erratics and glacial sediments are transported by icebergs. A good project is to photograph the icebergs one day, then to return the next day and rephotograph the lake from the same spot. Comparison of the photographs will show how much the seemingly static icebergs have changed in twenty-four hours. Icebergs in the lake are usually of the rectangular variety, which means that only about 1/7th of the iceberg is visible.

Local hikers have cut an unimproved, back-country hiker's style route along the edge of the mountain to the prominent point overlooking Valdez Glacier. Along the path hikers can spot striation marks on the black slate rocks, old moraines, and erratics as well as areas of glacial plucking and scouring. From the viewpoint, one sees the route of the '98s, the glacier's ablation moraine, morainal rocks, marginal crevasses in the glacier's ablation zone, old lateral moraines, and englacial and surface streams.

Today, Valdez Glacier is an increasingly popular winter and spring ski route for experienced skiers travelling up Shoup Glacier and down the Valdez or for those following the route of the '98ers up Valdez and down the Klutina.

Camica Glacier:

Camica Glacier was a tributary to Valdez Glacier until 1910. It is the prominent, hanging glacier in the valley across the lake from the parking area. It was named by Tarr and Martin for Dr. Camica, a Valdez optician and watchmaker who kept yearly records on Valdez Glacier at the turn of the century. Camica was born in northern Italy and sold as a small child to a band of wandering padrones or lazzaroni who travelled throughout France, Germany, England, Canada and Mexico. He escaped from slavery in Mexico and made his way to California where he worked days and went to school nights. At 56, he left California and went north with the '98ers. Until his death at 72, Camica served the people of Valdez and the many prospectors passing through. Camica was also an amateur glaciologist. As a hobby, he established observation stations near Valdez Glacier and kept yearly records on the advances and retreat of the glacier which he shared with both Grant and Higgins and later Tarr and Martin. Dr. Camica often told his friends that he wanted to leave his estate to the City of Valdez for a library, so that children growing up in Valdez could have the education he missed. Unfortunately, since he left no will, his estate was sold by the federal government and the proceeds went to the federal treasury under pre-territorial law (*Valdez Daily Prospector* May 13, 1912).

The Glaciers of Prince William Sound, Alaska

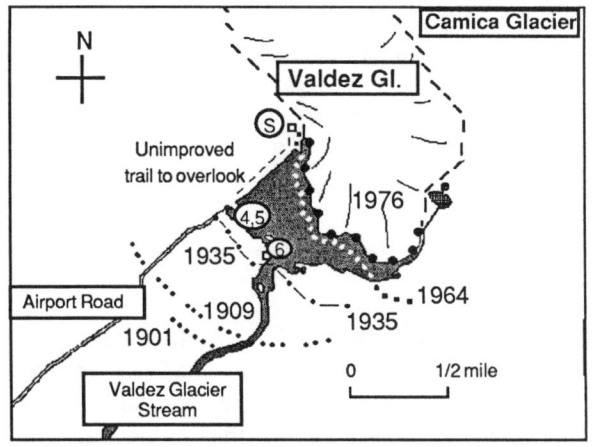

Fig-100. Positions of the Terminus of Valdez Glacier:

1901 -1986. Sources: Field, Dec. 26, 1985 unpublished map based on AGS parties in 1931, 1935, 1961, 1964, and 1976. 1909 Terminus estimated by Field from Tarr and Martin (1914) and 1901 terminus estimated by Field from measurements by residents of Valdez (written communication).

Photographic Stations:

All the early photographic stations had either become overgrown by vegetation or bulldozed for gravel by 1964. Field established new stations at the edge of the lake in 1964. Since then, Valdez Glacier has receded so much that it is barely visible from Stations 4 and 5. The best view is from station 6, which is adjacent to the Valdez Glacier stream. For experienced hikers, Photographic Station S(chrader) (Lethcoe) is on the prominent point overlooking Valdez Glacier Lake and the glacier.

Fig-101. Positions of the Terminus of Shoup Glacier.

Sources: Field, *Terminus of Shoup Glacier.*
USGS Valdez A-8 and A-7, Alaska, 1:63,360, 1960. (Based on 1957 aerial photography).

Post, aerial photographs 1954, 1965, 1966, 1976, 1984, 1986 (oblique).

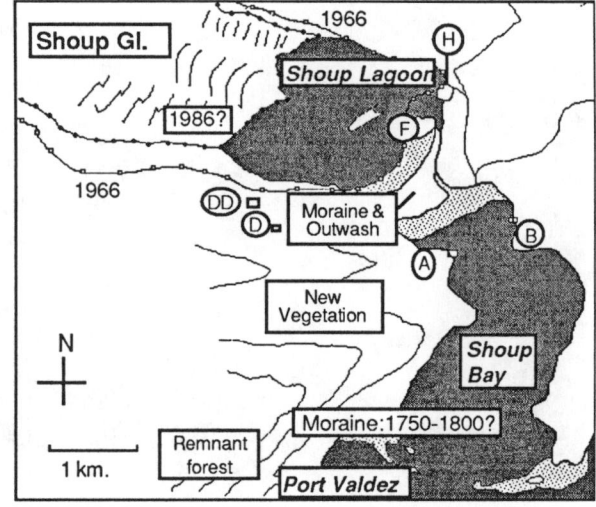

Photographic Stations:

Station A is the most historic site. It has been occupied since 1905. Station B is overgrown by alders. Stations D (Tarr and Martin) and DD (Field) are on a knoll at the 500 ft. level. They may be overgrown by alders. Field established photographic stations F and H on prominent knolls on the moraine. These are still good stations.

Shoup Glacier:

Location: Head of Shoup Bay, Port Valdez
Distance from Valdez: 7 miles
Access: Boat; cross-country hike from anchorage at Shoup
Type: Valley glacier descending from

Chugach Icefield
Length: 11.6 miles (Field 1975)
Area: 56.5 sq. miles (Field 1975)
Slope/aspect: Accumulation area SW; terminus SE
Status: Retreating
Photos: Fig-43, 102-105. Map: Fig-101

Glacial History:

Shoup Glacier is unique in the Prince William Sound area for having retreated from two terminal moraines and having scoured out two tidal basins. Shoup probably retreated from its outer morainal bar that abuts Port Valdez into the outer bay between 1750 and 1800. When Grant and Higgins visited Shoup in 1905 and 1909, they found it in slow

Fig-102. In 1961, Shoup Glacier's terminus sat on the terminal moraine it had occupied since Tarr and Martin first noticed the moraine's outer ledges appearing beneath the glacier in 1910, but it has retreated from the position occupied when first photographed by Washburn (Fig-43). Photo by Post, 1961. USGS.

retreat. The land between the terminus and the outer morainal bar was still mostly barren with patches of alder and a few spruce trees less than 3" in diameter.

Tarr and Martin (1914) were the first to report the presence of the second terminal moraine. They noted that the glacier's eastern ice cliff ended on an outwash delta while the western cliff was in shallow water. They also observed rock ledges showing beneath the glacier "at one point on the delta and two points along the tidal front." The appearance of bottom ice along the western front led them to predict that the glacier would soon cease to be tidal. Little did they know that behind the moraine was a another deep basin. After resting on this second moraine for over half a century, Shoup retreated from this moraine into a lagoon in the early 1960s. Recently glaciated areas around Shoup Lagoon are still barren. During the 1980s, Shoup Glacier remained grounded in deep water and continued to both shrink and retreat.

Fig-103: In the mid-1960s Shoup Glacier retreated off the moraine into deeper water. Black-legged kittiwakes have established a large rookery on the big island. Bathymetry studies show depths decreasing towards the face of the glacier. This possibly indicates that Shoup's retreat will slow as less of its face is exposed to saltwater. Photo by Austin Post, Aug. 29, 1984. USGS.

Fig-104. To transport supplies to gold mines on the ridges overlooking Shoup Glacier, miners constructed ice roads. Photo circa 1913. Mary E. Starr Collection. Courtesy of the Valdez Heritage Center.

Prospectors and Freighters on Shoup Glacier:

In 1907, after H.E. Ellis made his major gold strike at the Cliff Mine just outside of Shoup Bay, the Shoup area became the scene of intense mineral exploration. Claims were made throughout the area and several gold mines were developed near the terminus and on hillsides adjacent to the glacier's upper reaches. Miners freighted tons of supplies, including entire stamp mills, on roads carved out of the glacier (*Valdez Daily Prospector*, Aug. 11, 1913). In 1931, Field hiked up the glacier to the pass which overlooks Columbia Glacier. From there, one can see the nunatak which was the site of the Gold King mine. He reports that at that time "on the pass there were also remnants of a wire which must have been a telephone or telegraph line when the mine was in operation."

Shoup Glacier was also the scene of a number of deaths when avalanches descended on the miners. William Egan, the father of Alaska's first governor, died on Shoup Glacier when an avalanche buried his party (*Valdez Daily Prospector*, April 2, 1921).

In the early part of this century, Shoup Glacier was listed among the select group of glaciers with economic importance. Barges picked up the glacier's icebergs and sold them as ice blocks in Valdez and Port Liscum (Tarr and Martin 1914). Today, Shoup's economic importance lies in its value to the adventure travel market.

Fig-105. Miners using Shoup Glacier as a transportation route. Looking towards the Cameron-Johnson gold mine. Photo by P.S. Hunt. Circa 1913? Owen M. Meals Collection. Courtesy of the Valdez Heritage Center.

Points of Interest:

Shoup Bay is separated from Port Valdez by a morainal bar probably dating from the 18th century. Nautical chart No. 16700 reveals that Shoup Bay is a submerged hanging valley with respect to Port Valdez. Depths in Port Valdez reach 900 feet then rise rapidly to about 26 feet over the deepest part of the morainal bar. Boats with local knowledge can navigate the bar and anchor inside Shoup Bay. Depths in Shoup Bay reach to almost 300 feet. The 18th century trimline separating the older, conifer forest from the newer, alder shrublands with their scattered black cottonwoods can be seen on the prominent, low ridge on the western side of Shoup Bay. Vegetative covering of the slopes surrounding Shoup Bay still does not conceal the underlying roche moutonnee topography. Larger boats can anchor in Shoup Bay and either hike cross-country (rugged in places) or take a skiff up the river to Shoup Lagoon.

The second moraine forms a low-lying isthmus nearly landlocking Shoup Lagoon. The bedrock knolls reported by Tarr and Martin, a few kames and kettles, and glacial erratics are its major glacial features. This moraine is a major nesting area for Arctic terns, who build their nests on the ground. When crossing the moraine, it is best to walk carefully in a line so as not to inadvertently crush a nest or disturb nestlings. Land otter, bald-eagles, glaucous-winged gulls and mew gulls all feed on Arctic tern eggs and young. Predation can be significantly increased by careless human activity in a nesting area. The least intrusive places to walk are along the streambanks and in the intertidal zone.

Two streams cross the moraine. The smaller stream is shallow but can be navigated at high tide or portaged by kayaks. The northeast stream is the largest and best suited for power boats. At 0-tide it has a 10 knot current. Icebergs race through its narrow channel. At a +9 tide the current shifts direction, and water flows from Shoup Bay into the lagoon (personal communication, John Cotter). Icebergs may block the outflow. The deepest spot so far recorded in Shoup Lagoon is a little over 400 ft. Boats navigating the lagoon should be aware that large icefalls from the glacier have been observed to generate peaking and breaking waves at least 10 feet high. As at all calving glaciers, boats should not approach the glacier's terminus too closely. Falling seracs and bottom bergs may pose sudden hazards at any time. The absence of glacial activity is not an indicator of increased safety.

An island just inside the northeastern entrance to the lagoon provides prime nesting habitat for glaucous-winged gulls and black-legged kittiwakes. This is the largest and fastest growing black-legged kittiwake rookery in Prince William Sound. In 1987, the U.S. Fish and Wildlife Service biologists counted 3,359 nesting pairs and 2,499 chicks. Care should be taken not to disturb them.

Those on foot may walk from the inner spit around the southern side of the inner basin to the face of the glacier where there is a small area of kettle and kame topography. A series of prominent lateral moraines mark successive periods of shrinkage in the height of Shoup Glacier.

The rate of revegetation of newly exposed ground differs on the southern and northern sides of Shoup. Along the base of entire northern hillside, avalanches dump organic matter, seeds, and seedlings that quickly gain a foothold, since they do not have to compete with other plants for space, sunlight and nutrients. In 1987, two very large patches of ice, stranded by the glacier's retreat between 1984 and 1987, had young alders, willows, and herbaceous plants growing where avalanches had deposited them on the morainal blanket. The rate of revegetation in this area may be the fastest of any place in Prince William Sound. By contrast, the southern hillside has a gentle slope. No avalanches or rock slides bring organic material or seeds to the area. Bedrock surfaces are smooth and only sparsely covered by rock debris. Soil development is much slower. Seeds must be carried by wind, water, and animals. Early revegetation is proceeding much more slowly.

During the winter and spring, Shoup Glacier is a popular cross-country ski destination for either day skiing near the terminus or longer trips onto the icefield where one can look down on Columbia Glacier and across northeastern Prince William Sound. Both day trips and longer trips via the interconnecting icefields and glaciers are popular. Charter boat transportation and ski mountaineering guides are available in Valdez. From May to September, a tour operator provides interpretive day trips to Shoup Lagoon. Shoreside excursions with a naturalist are an important part of these trips.

Chapter 13. Drama at the Bridge: Glaciers of the Cordova Area

Between Valdez and Cordova, there are no tidewater glaciers. Several small apron and cirque glaciers cling to the peaks in Port Fidalgo, but it is not until one reaches Cordova and drives out the famed Copper River Highway that large glaciers are again encountered. Although there are no tidewater glaciers, Cordova area glaciers are intriguing because they are ice-calving glaciers which terminate in lakes or rivers.

Strictly speaking the Sheridan, Childs and Miles Glaciers are not located on Prince William Sound. But since Cordova, the point of access to them is on the sound, and because Childs and Miles Glaciers played such a dramatic role for a few months in the area's history, they are included here. All three glaciers plus several others can be seen from the Copper River Highway.

Sheridan Glacier:

Location: Mile 13.5 Copper River Highway
Access: road

Type: ice-calving glacier flowing from icefield
Length: 15 miles (Field 1975)
Area: 39 sq. miles (Field 1975)
Slope/aspect: SW/SSW
Status: retreating

Glacial History:

One of the most prominent features of the glacially modified delta in front of Sheridan Glacier is a series of concentric rings. These old terminal moraines are visible from the road or better from an airplane. According to Field (1975), the outermost ring, 3.1 miles from the glacier, represents the position of Sheridan's terminus during the late Pleistocene. An early Holocene retreat brought the terminus farther back into the mountains than at any more recent times. Sometime around 1700 Sheridan Glacier reached a late neoglacial maximum leaving a moraine that is inside the Pleistocene moraine. This is the outermost one of a close series of moraines between it and the present terminus. This series of moraines represents a fluctuating period of minor advances and retreats that continues to today. During its most recent advance in 1968, Sheridan pushed into the trees in some areas along its five and a half mile terminus. Today, it has retreated leaving a barren area of terminal lakes, push moraines and kettles scantily covered by invading plants. This recently deglaciated area in front of Sheridan Glacier is an excellent place to study the processes of revegetation.

An Observer's Guide to

Fig-106. Sheridan Glacier has a lobate terminus with lateral crevasses typical of valley glaciers which emerge on to a plain. Note small ice-dammed lakes (bottom) and kettles (wooded area). Photo by Austin Post, Sept. 12, 1986.

Tourism and Development:

The Chugach National Forest maintains a road to Sheridan Glacier and a campground in the adjacent woods. The Sheridan Mountain Trail, which starts from the campground, contours around the opposite side of the mountain from the glacier, so does not provide any glacial overlooks. Sheridan Glacier is a popular local destination for school and group outings.

To reach Sheridan Glacier, drive out the Copper River Highway about 13.5 miles (pavement ends at mile 12) until reaching a Forest Service road sign indicating the turn-off to Sheridan Glacier. It is 4.1 miles from the highway to the parking area along a well-maintained dirt road. Park in the open area near the Forest Service's Sheridan Mountain Trail and campground.

On foot, follow the rough four-wheel drive road from the parking lot down the hill to the rolling terminal moraine area. The raw, undercut slope, modified kame and kettle landscape, and old push moraines all indicate a recent slow retreat along the northwestern side. Combined lateral and ablation moraines cover the terminus. The best view point is from the top of a prominent push moraine dating from the 1968 advance that is unvegetated on one side and covered with alder on the other. From here, one overlooks the glacier's weathered seracs and light ablation moraine as well as the lake with its floating and grounded icebergs.

Childs Glacier:

Location: on Copper River
Distance from Cordova: 49 miles
Access: road, plane, raft or kayak

Type: ice-calving glacier flowing from icefield
Length: 12 miles
Area: 31 square miles
Slope/aspect: E/ESE
Status: advancing

Exploration:

Both Childs and Miles glaciers played important, nearly catastrophic roles in the exploration and development of the Copper River area. Childs Glacier was named by Lieut. Abercrombie during his 1884 summer attempt to ascend the Copper River. Prior to his arrival at the Copper River Delta, Abercrombie was under the impression that he would find the river navigable for 100 to 150 miles inland by steamer. He was quickly disillusioned.

> The water within a radius of about 30 miles of the mouth of Copper River is very dangerous for navigation. The glacial deposits from this river have formed innumerable bars and mud banks, visible only at extreme low water when the surf breaks over them. About 3:30 P.M. when about 5 miles from shore and about 8 miles from the northwestern mouth of Copper River, our oars suddenly came in contact with bottom. In another 10 feet our boat ran hard aground on a mud flat. Fortunately, there was little or no sea running or our boat, including its crew, would have been a total loss. (Abercrombie, 1884, p. 384.)

Abercrombie spent the summer trying to ascend the Copper River. His account of the events of July 13 are typical:

> On July 13 the reports caused by falling ice from the face of Childs Glacier (so called in honor of Mr. George Washington Childs,

of Philadelphia) and Miles Glacier (so called in honor of Brig. Gen. Nelson A. Miles) were terrific, and the ice floes on the river precluded all navigation. At this point the natives again begged to go back and were again bribed to help the expedition as far as the lower edge of Miles Glacier, and accordingly started across the river to its easterly connection. . . I steered, taking the water at a quartering angle, sometimes being swept back a mile or so only to try it again, and if the ice proved too thick or the current too swift, another channel was tried. (Abercrombie, 1884, p. 387).

By late August, Abercrombie had only gained the rapids above Miles Lake which bear his name. At this point, he finally conceded that the risks were too great to start a journey inland by foot without supplies and that a winter ascent, such as routinely followed by the Upper River Indians, was the only alternative. Lt. Allen returned a few months later in the winter of 1885 to make the first official ascent of the Copper River.

Recent Glacial History:

In his report on his 1884 attempt to ascend the Copper River, Abercrombie observed that "it is said by old Indians that the [Copper] river once ran under this glacier which filled the valley (Childs and Miles glaciers being united) and I am convinced that such was the case. . . "(p. 390). Glaciologists now believe that Childs and Miles probably coalesced during the neoglacial period and assign the prominent, tree-covered outer moraines, located on the far side of the Million Dollar Bridge, to this period.

On the basis of Abercrombie (1884) and Allen's (1885) reports on Childs Glacier, Tarr and Martin (1910) inferred that little change or possibly a slight recession occurred between 1884 and 1909. Then, suddenly, in 1909 to 1910 as engineers and workers rushed to complete the Million Dollar Bridge, the last vital link in the Copper River and Northwestern Railway which connected the wealthy Kennecott Mines at McCarthy to the new seaport of Cordova, Childs Glacier surged. In fourteen months, it advanced approximately 1640 ft. or over half the distance from its former position to the bridge. At the height of the surge from July 29 to August 6, 1910, Tarr and Martin believed that the middle of the glacier tongue was advancing at 130 feet a day. They wrote:

> What the glacier might do before the spring of 1911 was of great interest. It might continue to advance; or, as the diminishing rate of advance on the northern margin after August 11 suggested, it was possible that the strongest advance was over. If advance continued, would the glacier move up to and destroy the railway bridge, which was only 1571 feet distance from the northern margin, or would it stop before advancing that far? The bridge cost $1,500,000 and is the key to the new $20,000,000 railway to the copper mines.
>
> It is absolutely certain that no corps of engineeers living could save the bridge and railway if the glacier should advance that far. The

Fig-107. Childs (foreground) and Miles (background) glaciers both advanced when the Million Dollar Bridge was under construction. Recent flooding and erosion by the Copper River has changed the river bank opposite Childs Glacier. Photo by Austin Post, USGS, Sept. 3, 1966.

railway was less than a mile from the middle of the glacier, which might easily have advanced this distance between May and October 1910, when it was moving at the rate of 30 to 40 feet a day, or more. (Tarr and Martin, pp. 408-409)

To the great relief of the Guggenheims, who owned the copper mines and railway, and to all those working on the railway and expecting to live and work at the mines or in Cordova, the advance of Childs that had begun to slow in August continued to lose momentum over the winter. By the following summer it had diminished to a mere foot a day-

The bridge was safe. Childs began retreating in 1912 and continued until 1931. Then like many other glaciers in the Sound, it advanced for a period in the thirties. Field (1975) attributes this slow advance to climatic conditions. A retreat followed in the late '30s. The zenith occurred in 1959, when the glacier's ice cliff was at the lowest ever reco'ded. A very slow advance begun in the 1960s has continued into the 1980s.

Access to Childs and Miles Glaciers:

Childs and Miles Glaciers are reached by continuing out the Copper River Highway (unpaved) until reaching the Million Dollar Bridge at Mile 48. Park here and walk out on to the bridge for one of the most awesome sights in Alaska. To your left about one half mile away lies Childs Glacier. To your right, about five miles across a bend in the Copper River called Lake Miles looms Miles Glacier.

For the easiest viewing of Childs Glacier walk about 1/2 mile along a dirt road that turns off to the left just before reaching the Million Dollar Bridge to the former parking lot. Space in the parking lot is severely limited, because the Copper River has washed much of it away. Here, one looks straight across the river at Childs towering ice-cliff.

While watching for slabs and seracs to fall, one can observe how rock debris transported by the glacier has built a terminal moraine between the river and the glacier's terminus. From the parking lot, it is an easy 1/4 mile walk along the eroded river bank to a bend where one can see across to more of the glacier's terminus. It is not advisable to walk or stand near the edge of the river opposite the glacier. Blocks of falling ice can generate sudden large waves which may sweep along the bank with very little warning. Tarr and Martin give a dramatic account of these waves as they appeared during the 1909-1910 advance: ". . . the waves washed up over a bank 5 to 25 feet in height and rushed back 100 to 200 feet into the alder thicket. . . . Much gravel and sand, stones a foot or two in diameter, and ice blocks up to ten tons were thrown among the trees (Tarr and Martin, p. 408)." During this advance in 1909-1910, over 60 feet of the river bank opposite the glacier was eroded away by the increased current caused by the narrowing of the stream bed. Erosion of this river bank remains a critical problem. Severe flooding in 1983 and the erosive power of the river have washed away much of the Forest Service's parking lot. The Chugach National Forest plans to build new facilities.

To reach the land-terminating edge of Childs continue on across the Million Dollar Bridge, then take a steep trail that circles back under the bridge to the edge of the river. When the river is low, it is about a 3/4 mile walk over loose boulders to the glacier.

Another road, suitable for four wheeled drive pick-ups, leads off to the right just before reaching the Million Dollar Bridge and heads towards Miles Glacier. Local residents picnic on a point overlooking Miles Lake and Glacier. It is possible to find suitable spots for launching small boats. As with all ice-calving glaciers, boaters should exercise caution and maintain a healthy distance (1/2 mile) from the face of the glacier.

Bibliography of works referenced

Abercrombie, W.R. (1884), A Reconnaisance of the Copper River in 1884, In Explorations in Alaska. Twentieth Annual Report of the USGS to the Secretary of the Interior. 1898-1899. IV. Washington, D.C.: Government Printing Office.

____. *1899. Copper River exploring expedition.* Washington, D.C. Government Printing Office, 1900.

Burroughs, John. 1902. Narrative of the expedition. In *Alaska: giving the results of the Harriman Alaska Expedition carried out with the co-operation of the Washington Academy of Sciences.* I: 1-118. London: John Murray.

Calder, N. 1983. *Timescale.* New York: Viking Press.

Castener, Joseph C. 1898. *A Journey of hardship and suffering.* Ed. by Lyman L. Woodman. Cook Inlet Historical Society. Anchorage.

Cooper, W.S. 1942. Vegetation of the Prince William Sound region, Alaska, with a brief excursion into post-Pleistocene climate history. *Ecological Monograph.* 12:1-22.

Covey, C. 1984. The earth's orbit and the ice ages. *Scientific American.* 250(2):58-66.

Davidson. 1904. Glaciers of Alaska that are shown on Russian charts or mentioned in older narratives. *Trans. & Proc. Geog. Soc. Pacific.* 3:32-83.

Doroshin. Geology, Geognosy and Paleontology: From notes written in Russian America: 1848-1858. trans. unknown. Anchorage Consortium Library.

Field, William O. ed. 1975. *Mountain glaciers of the northern hemisphere.* 3 vols. Hanover, N.H.: Cold Regions Research and Engineering Lab, U.S. Army.

Environmental Protection Agency. 1979. Environmental Impact Statement: Alaska Petrochemical Company [ALPETCO], Refining and Petrochemical Facility: Valdez, Alaska. EPA-10-AK-VALDEZ-NPDES-79.

Fraser, A.B., and Mach. Mirages. *Scientific American.* 234(1):102-111.

Gilbert, G.K. 1910. Glaciers and glaciation. In *Harriman Alaska Expedition,* vol. III.

Glenn, E.F., and W.R. Abercrombie. 1899. Report of Captain Edwin F. Glenn on Explorations in Alaska. in *Reports of Explorations in the Territory of Alaska (Cooks Inlet, Susitna, Copper and Tanana Rivers) 1898.* Washington: Government Printing Office.

Grant, U.S., and D.F. Higgins. 1913. *Coastal glaciers of Prince William Sound and Kenai Peninsula.* USGS Bulletin 526.

Hazelet, George C. Hazelet Collection. Valdez Heritage Center. Alaska.

Heusser, C.J. 1955. Pollen profiles from Prince William Sound and Kenai Peninsula, Alaska. *Ecology* 30(2):185-202.

____. 1983. Holocene vegetation history of the Prince William Sound Region, South-Central Alaska. *Quaternary Research.* 19:337-355.

Icebergs and pack ice. *Encyclopedia Britannica.* 9:154-161. Chicago: Helen Hemingway Benton.

Jones, R.G. 1974. *Oceanographic investigation of Port Nellie Juan Fiord,* Alaska. Coast and Geodetic Survey.

Kalb, B. 1964. Glacier geophysics. *Science* 146(3642):353-365.

Kamb,B., C.F. Raymond, W.D. Harrison, H. Engelhardt, K.A. Echelmeyer, N. Humphrey, M.M. Brugman, T. Pfeffer. 1985. Glacial surge mechanism: 1982-1983 surge of Variegated Glacier, Alaska. *Science* 227 (4686):469-479.

LaBelle, J.C., Wise, Voelker, Schulze, and Wohl. 1983. *Alaska marine ice Atlas.* Arctic Information and Environmental Center. Anchorage.

Meier, M.F., Rasmussen, Post, Brown, Siknoia, Bindschadler, Mayo and trabant. 1980. *Predicted timing of the disintegration of the lower reach of Columbia Glacier, Alaska.* USGS Open-File Report 80-582.

Meier, M.F., Post, Krimmel, Driedger. *The 1983 recession of Columbia Glacier.* USGS Open-File Report 84-059.

Mendenhall, W.C. 1900. *A Reconnaissance from Resurrection Bay to the Tanana River, Alaska in 1898.* Twentieth Ann. Rept. USGS, 7:271-340.

Molnia, B.F. 1986. Glacial history of the northeastern Gulf of Alaska — a sysnthesis. In *Glaciation in Alaska.* eds. T.D. Hamilton, Reed, and Thorson. Alaska Geological Society.

Muir, J. 1902. The Pacific coast glaciers. In *Harriman Alaska Expedition,* I:119-135.

___. 1911. *Edward Henry Harriman.* Reprinted by Coastal Parks Assoc. 1978.

Nielsen, L.E. 1960. The Valdez and Klutina glaciers, Alaska. *Appalachia.* (June, 1960):31-37.

___. 1963. A glaciological reconnaisance of the Columbia glacier, Alaska. *Journal of the Arctic Institute of North America.* 16(2).

Post, A. 1975. *Preliminary hydrography and historic terminal changes of Columbia Glacier, Alaska:* USGS Hydrologic Investigations Atlas 559.

___. 1977. *Reported observations of icebergs from Columbia Glacier in Valdez Arm and Columbia Bay, Alaska, during the summer of 1976.* USGS Open-File Report 77-235.

___. 1978. *Interim bathymetry of Columbia Glacier and approaches, Alaska.* USGS Open-File Report 78-449.

___. 1980c. *Preliminary bathymetry of Blackstone Glacier, Alaska.* USGS Open-File Report 80-418.

___. 1980d. *Preliminary bathymetry of Harriman Fiord and neoglacial changes of Harriman, Surprise, and Serpentine Glaciers, Alaska.* Written communication.

___. 1980e. *Preliminary bathymetry of Upper College Fiord, Alaska.* Written communication.

___. 1980f. *Preliminary bathymetry of Barry Arm and Lower College Fiord and neoglacial changes of Barry and College Fiord Glaciers.* Written communication.

Post, A., and L.R. Mayo. 1971. *Glacier dammed lakes and outburst floods in Alaska.* USGS Hydrologic investigations Atlas HA-455.

Rasmussen, L.A., and M.F. Meier. 1982. *Continuity equation model of the drastic retreat of Columbia Glacier, Alaska.* USGS Professional Paper 1285-A.

[Remington, C.H.?]. 1939. *A golden cross(?): On trails from the Valdez Glacier.* By "Copper River Joe." Los Angeles: White-Thompson.

Rundquist, L.A. 1981. Valdez Flood Investigation. Anchorage: Woodward Clyde Consultants. Valdez City Hall Library.

Schrader, F.C. 1900. A Reconnaissance of a part of Prince William Sound and the Copper River District, Alaska, in 1898. In *Explorations in Alaska.* Twentieth Annual Report of the United States Geological Survey to the Secretary of the Interior. 1898-1899. IV. Washington, D.C.: Government Printing Office.

Shannon and Wilson, Inc. (1964). Subsurface Investigation for Mineral Creek Townsite: City of Valdez, Alaska. Seattle: Washington. Valdez City Hall Library.

Tarr, R.S., and L. Martin. 1914. *Alaska glacier studies of the National Geographic Society in the Yakutat Bay, Prince William Sound and Lower Copper River Regions.* The National Geographic Society.

Teben'kov, M.D. 1852. *Atlas of the northwest coasts of America from Bering Strait to Cape Corrientes and the Aleutian Islands with several sheets on the northeast coast of Asia.* Translated and edited by R.A. Pierce. Kingston, Ontario: The Limestone Press.

Vancouver, G. 1798. *A voyage of discovery to the North Pacific ocean and around the world in the years 1790--1795 . . . under the command of Capt. George Vancouver.* Ed. by John Vancouver.

Viereck, L.A. 1967. Botanical dating of recent glacial activity in western North America. In *Arctic and alpine environments,* edited by H.E. Wright, Jr., Osburn. Indiana UP.

Williams, John R., and K.M. Johnson, Surficial deposts of the Valdez quadrangle, Alaska. pp. B76-78. In *The United States Geological Survey in Alaska: Accomplishments during 1979,* eds. N.R.D. Albert and T. Hudson. Geological Survey Circular 823-B.

Williams, John R., and H.W. Coulter, Deglaciation and sea-level fluctuations in Port Valdez, Alaska. pp. B78-80. In *The United States Geological Survey in Alaska: Accomplishments during 1979,* eds. N.R.D. Albert and T. Hudson. Geological Survey Circular 823-B.

Index

Abercrombie, Capt. (Lt.), 50-51, 113, 128, 141,
Ablation zone, 8 (def.)
Accumulation Area Ratio (AAR), 15, 16,
Accumulation zone, 8 (def.)
Alaganik, 47
Alaskan Earthquake (1964), 29, 42, 80, 83, 90,
Alaska Railroad, 54, 60,
Albedo, 4, 9,
Altitude, 16,
Amherst Gl., 101,
Applegate, S., 50, 80, 81,
Apron Gl. 17, 49, 102, 126, 139,
Arête, 36, 37 (def.),
Aspect, 15, 100,

Baby Gl., 114,
Bagley Icefield, 17
Baker Gl., 85, 91,
Baltimore Gl., 98,
Barnard Gl., 98, 102-103,
Barry Arm 2, 15, 16, 25, 28, 34,
Barry Gl., 81-85, 86-87.
Batholith 3, 31
Bedload 36, 41 (def.)
Bergschrund 8, 22 (def.),
Bering Land Bridge (Beringa) 45,
Bettles Gl., 81,
Billings Gl., 62
Blackstone Bay, 64-65,
Blackstone Gl., 64-65
Borealis, Lake, 38
Bottomset beds, 42 (def.),
Brilliant Gl., 114,
Bryn Mawr Gl., 98, 107-108,

Burroughs, John, 86, 87,
Burns Gl., 59,

Calving 19, 24 (def.)
Camica, Dr., 132,
Camica, Gl., 124, 132,
Camica, Lake, 37,
Cascade Gl., 16, 23, 25, 81-85,
Castles, Lt., 105-106,
Cataract Gl., 24, 85, 87, 93
Chaos, Lake, 38,
Chatter marks, 31 (def.),
Chenega Gl., 16, 77-79,
Childs Gl., 139, 141-144,
Chugach Icefield, 17, 23, 99,
Chugach Mountains, 1, 9, 17, 56,
Cirque, 37 (def.),
Cirque glacier, 8, 49, 126, 139
Claremont Gl., 75,
Climate, 3, 4, 43, 49, 99,
College Fiord, 2, 4, 34, 35, 46, 99
Columbia Bay, 28, 35, 46,
Columbia Gl., 2, 16, 21, 23, 29, 34, 38, 54, 115-123,
Compressive flow, 23,
Contact, Gl., 18,
Cook, Capt. James, R.N., 19, 25, 116,
Copper River Basin, 46, 47
Copper River Delta, 42,
Copper River Hwy., 139
Cordova, 2, 139,
Coxe, Gl., 23, 34, 81-85,
Crescent Gl., 101,
Crevasses, 22, 25, 103,

Crossen, K., 55, 58,

Deformation, 10
Deltas, 42 (def.),
Detached Gl., 24, 85, 93,
Dirty Gl., 16, 96,
Doroshin, 50,
Downer Gl., 98,
Drift, 40 (def.),

Earthquakes, 3, 42 (cf. Alaskan Earthquake)
Eddy-holes, 32,
Erratic, glacial 41 (def.)
Esker, 41 (def.)

Falling Gl., 26, 76,
Fidalgo, Lt. Salvador, 111, 117,
Fidalgo, Port, 33
Field, William O., 53-54, 52, 62, 63, 65, 66, 68, 72, 73, 74, 75, 76, 77, 78, 79, 80, 81, 85, 90, 92, 93, 94, 96, 98, 102, 104, 106, 107, 108, 110, 111, 112, 114, 122-123, 133, 134, 141, 144,
Fiord, 33, 35-36 (def.)
Firn, 9 (def.), 10
Firn limit 8, 9 (def.)
Foliation, 11 (def.)
Foreset beds, 42,

Gannett H., 87,
Gilbert, G.K., 51, 105-106,
Gilbert, Mt., 89,
Glacial erratic, 41 (def.)
Glacial ice, 5, 10
Glacial movement, 13
Glacial streams 24,

149 / Index

Glacial surges 14,
Glacier-dammed lakes, 34, 37, 38, 126-127,
Glacier, types
　Apron, 17
　Calving (Tidewater), 5, 18-21, 118,
　Cirque, 17,
　Hanging, 33,
　Polar, 5, 17
　Stairstepping, 23,
　Temperate, 5, 17
　Tributary, 27,
　Valley, 18,
Glenn, Lt., 81, 101, 113,
Golden, 46, 49,
Gouge holes, 32
Grant, U.S., 52,
Grant, U.S. and Higgins, D.F., 64, 66, 67, 73, 76, 79, 85, 93, 94, 98, 107, 113,-114, 126,
Gravina, Port, 33
Great Nunatak, 118, 123,

Hanging Valley, 33, 34 (def.), 35
Harriman, Edward H., 51, 86,
Harriman Alaska Expedition, 51-53, 83, 93, 101, 117,
Harriman Fiord, 2, 15, 16, 86-92, 95-97,
Harriman Gl., 16, 85, 96-97,
Harvard Gl., 16, 98, 109-110,
Hazelet, George C., 129-131,
Heather I., 117, 118, 119, 120-121,122,
Heusser, Calvin, 48, 49, 55,

Higgins, D. F., 51, 113,
Hinchinbrook I., 44, 45, 46,
Holyoke, Gl., 25, 98, 102-103,
Holocene glaciation, 44, 46,
Horn, 33, 36, 37 (def.),

Ice avalanches, 23-24, 93,
Ice, bubbles, 10
Icebergs, 11-13, 39,
Ice-cones, 26,
Ice-crystals, 10, 11,
Ice-dammed lakes, 34, 37-38, 126-127
Icefalls, 23
Icefields, 17-18 (def.),
Icy Bay, 2, 28, 77,

Kame, 40, 41 (def.)
Kadin, Lake, 38, 115-116, 117,
Kamb, B., 55,
Keen, Dora, 83,
Kenai Mountains, 56,
Kettle, 41 (def.),
Keystone Canyon, 37, 38,
Kinematic wave, 14,
Kings Bay, 50, 73,
Knight I., 44, 46,

Lakes, glacier-dammed, 34, 37-38, 126-127

Maps, 50-54,
Meares Gl., 16, 27, 111-114,
Meier, Mark, 19, 55,
Mendenhall, W.C., 51, 58, 105,
Miles Gl., 2, 14, 143-144,
Mineral Creek Delta, 42, 47, 48, 125,

Mirage, 80
Miocene glaciation, 43-44,
Montague I., 44, 46,
Moulin, 24, 41,
Morainal rocks 28, 29,
Moraines,
　ablation, 8, 28,
　emergent, 26, 92
　lateral 24-25 (def.), 33, 88, 90, 105,
　medial, 26, 33,
　push, 28
　terminal, 8, 27, 28,
Muir, John, 4, 51, 86,
Muir, Mt., 89, 93,

Nanatuk Lakes, 115-116, 118, 122-123,
Nellie Juan Glacier, 7, 9, 16, 18, 23, 31, 37, 41, 67-72, 121,
Nellie Juan, Port, 2, 67,
Nielsen, L., 5, 131,
Number One, Lake, 38,
Nunatak, 23, 36 (def.)

Outwash, 40 (def.)
Outwash plain, 40, 41 (def.)

Passage Canal, 33, 34, 56,
Penniman Gl., 90,
Perry I., 46,
Photographic sta., 6 (def.)
Pigot Bay, 35,
Pigot Gl., 80,
Pleistocene glaciation, 35, 44, 56, 124,
Pliocene glaciation, 35, 44,
Plucking, 30 (def.)
Plunge-pool potholes, 32
Port Wells, 80,
Portage Gl., 57-61,
Portage Lake, 58,

Post, Austin, 15, 16, 19, 54, 64-65, 67, 80, 115, 72, 73, 84, 91, 92, 94, 103, 106, 109, 110, 115, 118, 119, 122-123, 125, 126-127, 133,
Post, Austin (photos), 9, 18, 21, 23, 26, 38, 69, 70, 107, 109, 112, 134, 135, 140, 143,
Precipitation, 3,

Prince William Sound, 2, 43 (geology of); 3 (climate); 35 (fiord),
Princeton Gl., 18,

Radcliffe Gl., 98, 109,
Radiocarbon dates, 46-48, 95, 97, 111, 115, 124,
Ranney Gl., 114
Revegetation, 49, 121, 138,
Roche moutonnee, 37 (def.)
Roaring Gl., 24, 85, 95,

Salt-marshes, 42
Sandur, 42 (def.)
Sargent Icefield, 17, 18, 56,
Schrader, F.C., 51, 129,
Scouring, 30,
Sedimentation, 35, 36, 41,
Seracs, 23 (def.)
Serpentine Cove, 90,
Serpentine Gl., 22, 85, 87, 88, 89-90,
Seth Gl., 3, 63,
Shearlines, 30,
Sheep Creek, 38, 42,
Sheridan Gl., 2, 139-141,
Sherman Gl., 29,
Shoup Bay, 37, 133-138,
Shoup Gl., 25, 53,
Smith Gl., 98, 108,

Snowfall, 5, 14,
Snowline, 9
Strandlines, 38 (def.)
Streams, glacial, 24, 32, 41-42,
Striation 31 (def.),
Surprise Gl., 15, 16, 93, 94-95, 85,
Surprise Inlet, 92, 93,
Suspension load, 36, 41 (def.)

Tarr and Martin, 57, 61, 62, 63, 90, 99, 102, 104, 108, 110, 126, 134, 135, 142, 144
Tarn, 37 (def.),
Taylor Gl., 41, 73
Tebenkof Gl., 15,18, 63,
Teben'kof, Capt. M.D., 63, 117,
Terentiev, Lake, 38, 115-116, 117, 123
Tidewater glacier, 19-21,
Tiger Gl., 16, 79
Tigertail Gl., 18,78
Till, 39-40 (def.)
Thompson Pass, 31,
Toboggan Gl., 16, 85, 91,
Topset beds, 42 (def.)
Trimline, 33, 34 (def.), 83,
Truncated ridge, 33, 35 (def.).

Ultramarine Gl., 18, 67,
Unakwik Inlet, 2, 28, 34, 35, 36,
U-shaped valleys, 33,

Valdez, 2, 42, 48-49, 124-126,
Valdez-Fairbanks Trail, 32
Valdez Gl., 26, 31, 37, 51, 54, 125, 126-133,

Valdez Glacier Delta, 42,
Valdez Glacier Route, 127-131,
Valdez Glacier Valley, 33, 49,
Valdez, Port, 33, 34, 36, 41, 46, 47, 124,
Valley, 33, 34,
Vancouver, Capt. George, R.N., 19, 50, 57, 77, 81, 100, 112,
Vassar Gl., 98, 104,
Viereck, L.A. 55,
Volcanic ash, 9-10

Washburn, Bradford, 53, 54
Wedge Gl., 16, 85, 96,
Wellesley Gl., 23, 98, 103,
Wellesley Cove, 35,
Wells, Port, 34, 35,
Whittier, 2, 56,
Whittier Gl., 61-62,
Whittier Icefield, 17, 56,
Wisconsin Age glaciation, 45-48,
Worthington Gl., 25,

Yakataga Rock Formation, 44,
Yale Arm, 35,
Yale Gl., 16, 98, 105-107,